DARK COSMOS

IN SEARCH OF OUR UNIVERSE'S MISSING MASS AND ENERGY

DAN HOOPER

Smithsonian Books

Collins

An Imprint of HarperCollinsPublishers

All illustrations appearing throughout the text are copyright © 2006 by Patricia Wynne, except: p. 14 (right)—courtesy of Carnegie Institute of Washington; p. 14 (left) and p. 33—courtesy of California Institute of Technology; p. 23—courtesy of H. Bond, et al., Hubble Heritage Team (STScl/AURA); p. 74 © SPL/PhotoResearchers, Inc.; p. 109 © IceCube Collaboration; p.152—reprinted with permission of Lucent Technologies; p. 154 © NASA/WMAP Science Team.

The poem appearing on page 63 is reprinted from *Collected Poems 1953–1993* by John Updike. Used by permission of Alfred A. Knopf, a division of Random House, Inc.

HarperCollins books may be purchased for educational, business, or sales promotional use. For information please write: Special Markets Department, HarperCollins Publishers, 10 East 53rd Street, New York, NY 10022.

First Smithsonian Books paperback edition published 2007

Designed by Daniel Lagin

The Library of Congress has catalogued the hardcover edition as follows:
   Library of Congress Cataloging-in-Publication Data
      Hooper, Dan, 1976-
         Dark cosmos: in search of our universe's missing mass and and energy/Dan Hooper.
            p. cm.
         Incluides bibliographical references and index.
         ISBN 978-0-06-113032-8
         1. Cosmology—Popular works. 2. Physics—Popular works. I. Title.

      QB982.H66 2006
      523.1—dc22

                                                                          2006044333

ISBN 978-0-06-113033-5 (pbk.)

07 08 09 10 11   ID/RRD   10 9 8 7 6 5 4 3 2 1

CONTENTS

INTRODUCTION

I t is a common misconception that scientists want everything to be neat and tidy—they don't want anyone questioning their ideas, or suggesting that current explanations are incomplete. This perception is, of course, completely off the mark. After all, if we scientists knew everything, there would be nothing left to discover! It is not completely facetious, however, to suggest that what every scientist dreams of doing each day is to prove his or her colleagues wrong.

What scientists find most exciting are mysteries—puzzles with the final pieces not yet in place. And this is why modern cosmology has captured the attention of so many physicists, theorists, and experimentalists from many different subfields. Few mysteries in nature are as deep, or as puzzling, as those associated with observational discoveries in cosmology over the past thirty years or so.

Almost twenty years ago, I wrote that we are in the midst of perhaps the greatest Copernican-like revolution in history. Copernicus, you will recall, boldly and correctly argued that Earth was not the center of the solar system. In the 350 years or so since his insights were confirmed, humans have, if anything, continued to be relegated farther and farther from the center of the Universe. Not only is Earth not the center of the solar system, our solar system is not the center of our galaxy, instead residing in a boring suburb on its outskirts. And our galaxy is not the center of our cluster of galaxies, and so on. Indeed, there is absolutely nothing special about our location in a Universe that we find is more or less the same in all directions.

Things are even worse. Not only are we not at the center of the Universe, but we have now discovered that if you got rid of us, our planet, our Sun, all the stars and galaxies and gas that we can see through our telescopes, the Universe would still be largely the same! We are, if anything, a small bit of pollution in a Universe dominated not by matter such as that which makes us up, but rather matter and energy that appear to be completely different from anything we have ever observed on Earth. If we did not feel cosmically insignificant before, we should now.

We have learned that the dominant material in our galaxy, and pretty well all galaxies we can see, is invisible to telescopes. It doesn't shine. The good news is that if this odd sort of material is made from some new type of elementary particle, then the "dark matter" that dominates our galaxy is not just "out there," it is in the room as you read this, traveling through the paper, and through your body. In this case, if we are clever, we can design experiments here on Earth to detect this stuff!

What makes the discovery of dark matter even more exciting is that our study of the fundamental structure of matter, called elementary particle physics, is at a crossroads. The current "standard model" of particle physics, which thus far has explained everything we can see, is nevertheless incomplete. Hidden behind the mathematical predications that agree so well with observations are deep dilemmas. We hope to resolve some of these dilemmas through the next large particle accelerator, called the Large Hadron Collider, due to come online in 2008.

Of course, theorists have not been idle while we wait for this machine to turn on. We have been developing sophisticated guesses as to what nature may reveal as we open this new window on the Universe. And guess what? Almost all these guesses involve new particles and fields—thus far undetected—that could be the mysterious dark matter we seem immersed in today.

In this sense, then, the mystery of dark matter ties together the major outstanding puzzles in two complementary and seemingly disparate fields, particle physics and cosmology—the physics of the very small and the very large, respectively. Resolving this mystery might therefore produce not just one, but two quantum leaps in our understanding of nature.

This alone would be reason enough to celebrate the emerging field of particle astrophysics. But almost a decade ago, an even more puzzling discovery made the idea of dark matter, as exotic as it is, seem tame.

We have discovered that by far the biggest form of energy in the Universe does not involve matter at all! Most of the energy in the Universe seems instead to reside in empty space. This energy is gravitationally repulsive and appears to be causing the speeding of the observed expansion of the Universe.

To suggest that we don't understand much about dark energy is an understatement. Why empty space should have the energy it apparently does is probably the biggest mystery in all of physics. The recognition that this energy appears to exist has completely altered the landscape of theoretical particle physics, while at the same time driving a host of astronomers and astrophysicists to launch new cosmic probes to try and discern its nature.

What we do know is that whatever the nature of this energy,

its origin probably is related to the origin of our own Universe, and its future will guide the future of our Universe. For these reasons, we cannot ultimately answer the questions asked by humans since they first started to think—i.e., "Where did we come from?" and "Where are we going?"—until we understand the nature of this "dark energy," as it has become known.

The search to understand the nature of dark matter and dark energy is perhaps the grandest adventure we have ever undertaken. It involves the most sophisticated technological devices humans have ever built— from large accelerators to large telescopes, from sensitive devices built underground to ingenious satellites we launch into space. It is a story worth telling, and a story worth reading about for anyone who has ever looked up at the night sky with awe and wonder.

—Lawrence M. Krauss
Cleveland, Ohio, 2006

When I first entered the world of higher education as a college freshman, I never imagined that four years later I would be leaving that institution on my way to becoming a professional physicist. In fact, I had enrolled with the idea of majoring in music. My loud and distorted electric guitar playing didn't impress the music faculty very much, however, and the Bach concertos they encouraged me to study never really spoke to me. After a few months, I began looking for a new major.

Over the next year or so, I decided to become an economics major, then a history major, then pre-law, then economics again, then business or maybe finance, and then engineering (probably electrical, but I hadn't decided). Still, after all of this intellectual bouncing around, I hadn't really found anything that excited me. All of these subjects had their interesting moments, but as much as I wanted to be, I just wasn't enthralled by any of them. At least an engineering major would get me a good job, I thought. And I was good enough at math to get through the courses without too much trouble.

Thankfully, my story does not end with me designing software for Microsoft or IBM, but instead led me to something that I find much more interesting. The engineering curriculum required me to take a year of introductory physics and a course or two on modern physics. I muddled my way through that first year, dragging my feet as I went, and expected to make it through the next year in a similar way. On the first day (or maybe first week, I can't recall) of my required modern physics class all of this changed. It was there that I heard about quantum physics for the first time. Despite what I had always thought up until that point, the Universe wasn't boring at all. It was crazy and amazing! It was completely different from what I—or anyone else—had expected it to be like. And it wasn't only quantum physics that was so strange. Later in that course, I learned about relativity for the first time. This was all completely mind-blowing stuff—and I wanted more.

I started asking the physics faculty questions that went beyond the scope of the second-year physics curriculum. Most of them were happy to answer my questions, but I don't think I understood very much of what they had to say, and I was certainly too impatient to wait a few more years until I had taken the courses I needed to follow their explanations. Fortunately, one of my professors loaned me a copy of a book written by Paul Davies called *Superstrings*. I've said on many occasions that I became a theoretical physicist because of that book.

One thing that people who know me understand is that I have a highly obsessive personality. After reading *Superstrings,* I felt as addicted as any heroin junkie. In the next year or so, I read five or ten other Paul Davies books along with other popular physics books by Michio Kaku, Kip Thorne, Richard Feynman,

John Gribbin, and others. The trajectory of my life was set in motion. I had no other choice but to become a physicist.

Unlike many of my professional colleagues, I still read popular physics books. I don't read them to learn new things about physics anymore, however. I read them for inspiration. It is easy to forget how exciting and incredible modern science truly is. Scientific articles found in academic journals very rarely capture the sense of wonder and awe that originally motivated me to become a physicist. Nevertheless, under all the layers of mathematics and terminology, the ideas contained within many of those articles *are* wondrous and awesome.

My hope is that this book you are about to read captures some of the amazement that I feel about physics and cosmology. I still remember how those first popular science books once felt. It is in this spirit that I have tried to convey in this book the ideas and discoveries that I consider to be among the most exciting of modern physics. I hope that I have managed to capture some of that fascination that I remember experiencing for the first time.

I got a great deal of help from my friends, family, and colleagues in writing this book. I would like to thank Jodi Cooley, Gerry Cooper, Kyle Cranmer, Jon Edge, Josh Friess, Antony Harwood, Becky Hooper, Lori Korte, Jo Rawicz, Constantinos Skordis, Andrew Taylor, Roberto Trotta, John Wiedenhoeft, and whomever I am forgetting (I can guarantee that there are others who deserve to be mentioned here) for their advice, comments, and proofreading. I would also like to especially thank my editor T. J. Kelleher, who has been invaluable in turning a rough collection of words into what I hope is an enjoyable book.

DARK COSMOS

CHAPTER 1

OUR DARK UNIVERSE

The world is full of obvious things which nobody
by any chance ever observes.

—Sherlock Holmes

Take a look around you. You see a world full of things. Tables, chairs, the floor, a cup of coffee, shoes, bicycles—things. Most of us casually think of the world as space filled with such things, the sort of stuff you can hold in your hand or stub your toe on. But how much of our world is really made up of objects that you can see? Think of the air you're breathing. It's invisible. Nevertheless, it is there, even if your experience of it is somewhat indirect as your chest expands and contracts, and your breath whistles through your nose. The visible world is not all there is to the Universe. Relying solely on our eyes to learn what's out there would cause us to overlook a great deal.

Although the point I'm making might seem obvious, it is one worth bearing in mind. Just as we cannot see the air, we cannot see most of the Universe. During the past several decades, several lines of evidence have led to the conclusion that about 95 percent of our Universe's mass and energy exists in some form that is invisible to us. Hidden. Evading our detection almost entirely. That might seem ridiculous, but just as the act of blowing up a balloon helps us see the air we breathe, our hidden Universe does leave clues that we can decipher to confirm its existence. Galaxies are seen rotating at much greater speeds than are possible without the presence of extra matter. And the large-scale structure and evolution of our Universe, from the Big Bang to the present-day expansion and acceleration, seem to require more mass and energy than we see—some twenty times more. This picture of the invisible gets weirder. Of this mysterious and subtle majority of our world, only about a third is thought to be matter. Appropriately, it is called dark matter. The other two-thirds is stranger yet, and is called dark energy.

Thousands of physicists, astronomers, and engineers are actively working toward the goal of understanding the nature of dark matter and dark energy. Many of these scientists are skilled experimenters, designing ultra-sensitive detectors in deep underground mines, constructing new kinds of telescopes capable of detecting much more than simply light, or operating particle colliders that smash matter together at incredible speeds. Others, such as I, are theoretical physicists, struggling to understand with pencil, paper, and powerful computers how dark matter and dark energy fit into our world as we currently understand it.

Although the scope of these collective efforts is staggering, the basic motivations are nothing new. For as long as people have pondered their world, they have tried to identify what it is made of. The philosophers of ancient civilizations speculated and hypothesized endlessly on such matters, if not always very successfully. Millennia later, but still in much the same spirit, the heirs to those philosophers discovered and codified the chemical elements of the periodic table that we are all taught in school. Twentieth-century physics has further revealed an incredible world of quantum particles. These particles are part of a beautiful and elegant theory that successfully describes nearly all of the phenomena observed in our Universe. But, alas, nearly all is not nearly enough.

• • •

Long before the advent of modern chemistry and physics, the peoples of early civilizations made countless attempts at understanding the composition of the things around them. The ancient Greek philosopher Empedocles provided one of the most enduring of those ideas when he hypothesized that each type of matter in the Universe arises from a specific combination of four fundamental elements: air, earth, fire, and water. Empedocles, followed by Plato and a long list of others, thought that it would be impossible to change one pure element into another, but by melding together different quantities of these pure elements, any substance could be formed.

The healthy system of discussion and debate among learned Greeks fostered further investigation. Elementalists, such as the philosopher Democritus, conjectured that all matter was made up of a finite number of individual, indivisible particles that he called atoms. Democritus believed, as did the other elementalists, that these fundamental particles could not be destroyed or created, but only arranged in different patterns or in different quantities to make different substances. A slippery substance, for example, would be made out of round, smooth atoms. An object made up of atoms with hooks or other such shapes could stick or lock together in dense groups to form heavy substances, such as gold. This basic idea of Democritus's turned out to be, very roughly, correct.

Modern chemists know that the qualities of a substance are not so simply determined by the superficial properties of atoms themselves, but instead largely result from the interactions among atoms. But despite the failure of the ancient elementalists to build an accurate atomic theory, the concepts at the foundation of their theory represented a major step forward in scientific thought. Many of the concepts are essentially the same as those taught in nearly every chemistry classroom today. The atoms of modern chemistry, however, are not the indivisible and fundamental objects envisioned by Democritus.

During the twentieth century, as experimenters probed deeper into the nature of the atom, they found that atoms are not indivisible. Experiments by physicists such as J. J. Thomson and Ernest Rutherford showed that atoms themselves are made up of constituent parts: protons, neutrons, and electrons. And in a further refutation of Democritus, physicists found that one element could be changed into another by adding or removing those parts. Modern-day alchemy—but without the appeal of gold. In the 1960s and 1970s it was learned that protons and neutrons themselves are made up of even smaller particles. It seems that the Greek concept of the atom applies more to these smaller particles than to the objects in the periodic table that we call atoms.

• • •

Well, so what? When scientists discover and catalog the building blocks of our world, what are they really accomplishing? At a minimum, a discovery is something with value in itself. Whether it be a discovery of a new species of bird, a new planet, or a new type of elementary particle, knowing of its existence tells us something about our world. Throughout most of scientific history, these kinds of accomplishments were the main goal of scientists.

Botanists made lists and sub-lists of the species of plants they knew to exist. Early chemists cataloged the known types of metals, gases, and other substances. Astronomers discovered ever increasing numbers of stars, comets, planets, and moons.

Occasionally, however, the quest for discovery can reveal something even greater. Biology, for example, was dominated by cataloging until the nineteenth century, as zoologists and botanists generated lists of species and categorized them by their characteristics, work exemplified by that of Carolus Linnaeus. But when Charles Darwin developed his theory of biological evolution through the process of natural selection, he did much more than generate a new list of species—he explained why the lists, drawn up by Linnaeus and others, were the way they were.

Modern physicists hope, like Darwin, to find not only a more complete description of nature, but also a more complete explanation for it. The discovery of new types of particles in our world is often seen as only a step toward that goal. Without an accompanying explanation for why some particle exists, such discoveries leave most physicists dissatisfied. The community of particle physicists has taken upon itself a quest to find the ultimate explanation. Much like Darwin, we are in search of a reason for why things are the way they are.

The great nuclear physicist Ernest Rutherford famously said, "In science there is only physics; all the rest is stamp collecting." Although this was a particularly harsh choice of language, Rutherford hit upon an important distinction. By stamp collecting, Rutherford meant something similar to what I call list-making. The "physics" he refers to, on the other hand, is not asking questions of what, but questions of why.

Regardless of what Rutherford might have to say about it, so-called stamp collecting is vital to the advancement of science. Consider chemistry. Nineteenth- and early twentieth-century chemists had empirically identified enough elements as well as enough structure and pattern in their characteristics to group them into the modern periodic table, which is essentially a list of substances and their characteristics. Following the discovery of the neutron and the realization that all of the atoms of the periodic table were combinations of protons, neutrons, and electrons, however, an organizing principle emerged. With this realization, it became possible to determine what kinds of elements could exist by considering different ways in which protons, neutrons, and electrons could be bound together. Even elements that hadn't been discovered yet could be reliably predicted to exist.[1] The theory behind the table even explains why each element has the attributes it does. Discovering new atomic elements became a means to confirm the predictions of modern atomic theory, rather than an end in itself. Cataloging elements had evolved into something that even Rutherford would be proud to call physics.

In recent years a new list of characteristics of the Universe has been drawn up. Like the lists I've just discussed, the new list cries out for an explanation. In the remainder of this chapter and in the following ones, I will tell many of the stories behind the making of this list, and the overarching story of how many scientists came to the conclusion that decades' worth of observations are best explained by the existence of dark matter and dark energy. That conclusion has driven numerous physicists and astronomers to take up the formidable quest of uncovering the natures of these elusive substances. This quest is not merely to add descriptions of dark matter and dark energy to our list of constituents of the Uni-

---

1. Even before protons, neutrons, and electrons were known to exist, hints of this organizing principle had been noticed. Dmitry Mendeleyev, who originally drew up the periodic table, unwittingly used these patterns to predict correctly the existence and characteristics of elements, such as germanium and gallium, that had not yet been discovered.

verse, but to better explain and understand why our world is the way it is.

. . .

Physicists first got an inkling that there was more to the Universe than, well, meets the eye when astronomers, seeking to catalog everything in space, applied the lessons of physics learned here on Earth to the heavens. Such a program was possible because atoms and molecules, whether in a laboratory or the farthest reaches of space, radiate light. Radiation occurs when electrons, hovering in clouds surrounding the nucleus of an atom, absorb and release energy. The released energy escapes in the form of particles of light called photons. The light produced does not have random attributes: electrons in a given type of atom or molecule release light in frequencies specific to that atom or compound. Those frequencies correspond to a specific spectrum of light for each atom or compound—copper's spectrum, for example, is largely green, and table salt's is yellow-orange. If you observe a specific spectrum of light, then you can identify the chemical makeup of the material releasing it. With this tool, known as spectroscopy, astronomers and physicists are able to identify the atoms and molecules that make up the Sun, the planets within our solar system, interstellar clouds of dust and gas, distant galaxies, and most everything else there is to see.

Using this tool, astronomers began to catalog the substances that fill every corner of our Universe. As they collected more and more data in this way, they found that Earth is not special, at least not in terms of its chemical composition. The same chemical elements and molecules that were cataloged by chemists on Earth were found in stars and clouds of dust and gas everywhere in the Universe astronomers looked. Although the quantities of various elements varied from place to place, it was all the same stuff. The chemical world appeared to be more or less the same everywhere. But it became apparent soon enough that there was much they couldn't see.

What kind of object could escape the watchful eyes of our telescopes? To begin with a simple example, consider planets like Earth or the others in our solar system. We expect that planets should exist not just in our solar system, but more or less throughout the Universe. Confirming the existence of planets outside of our solar system isn't easy, however. Unlike stars, planets don't give off much light, and are thus incredibly difficult to detect using the usual methods of astronomers. Today, more than one hundred very heavy planets (sometimes called Jupiters because they are similar in mass) have been observed in other solar systems, but often not by observing light from them. Instead, they are detected indirectly by observing whatever star they orbit. When a very heavy planet orbits a star, it causes the star itself to move. Although an astronomer may not be able to see a planet itself, its presence can be inferred by watching a star wobbling back and forth. For a smaller planet like one the size of Earth in a distant solar system, the observational methods are just now being developed. We can only speculate as to how many planets like ours might exist.

All of this being said, the total mass of the planets surrounding a given star is almost certainly going to be considerably less than the mass of the star itself. All of the planets in our solar system put together, for example, are only about 0.1 percent as massive as the Sun. So the total mass of planets is not likely to be a very large fraction of the total Universe. Nevertheless, the case of invisible planets does illustrate a point—matter might exist that has remained invisible to our experiments. If there are planets that we cannot easily see, there could easily be other objects or substances just as difficult—or even more difficult—for us to detect.

Such stuff needn't be planets, however. So far we have been talking only about some of the most ordinary types of matter: objects made up of protons, neutrons, and electrons. Throughout the course of the twentieth century, physicists have discovered dozens of new elementary particles, many of which have properties radically different from those found in ordinary matter. Although most of those entities exist for only a fraction of a second before disintegrating and leaving behind other particles in their place, some of them are stable. They are out in space, and all around us. You can forgive yourself for not noticing—for all their ubiquity, they are exceedingly difficult to see.

Great, you might say: if such wild particles do exist, how do we know? Well, recall how those massive Jupiter-like planets were observed—through the effects of their gravity on stars. Perhaps gravity or one of the other three known forces of nature—the electromagnetic force, the strong nuclear force, or the weak nuclear force—can help. All known particles interact through some combination of those forces. Gravity, for example, affects anything with mass or energy, which is everything. In addition to gravity, electrons feel the electromagnetic and weak nuclear forces. Protons and neutrons interact through all four known forces. It is the strong nuclear force that holds quarks together to form protons and neutrons, and holds protons and neutrons together inside of atoms in a compact nucleus (we'll revisit that later).

The electromagnetic and strong nuclear forces are the two strongest forces of nature. It is these forces that make protons, neutrons, and electrons easy to detect and observe. The reason your hand stops when you press it against a table is that the electric charge inside the table repels the electric charge in your hand. This is the result of the electromagnetic force. If this force were somehow turned off, your hand would simply pass through the table.[2] Although these powerful forces make protons, neutrons, and electrons easily detectable, it is natural to wonder if particles not affected by these forces might also inhabit our Universe, evading our most careful observations.

Imagine a particle, for example, that is not affected by either the strong nuclear or electromagnetic force but only by gravity and the weak nuclear force. Such a particle exists, and we call it the neutrino. Neutrinos are very light and fast, and because they do not feel the strong nuclear or electromagnetic force, they scarcely interact with ordinary matter. Neutrinos can often travel through the entire Earth without interacting with it or anything on Earth, such as telescopes. There are hundreds of billions of neutrinos traveling through your body every second of every day, entirely undetected, entirely invisible.

The neutrino's elusiveness leads to a question crucial to understanding the composition of our world. If there were enough neutrinos in the Universe, could they outweigh all of the more familiar stuff? Could there be more total mass in neutrinos than in atoms? Could there be more mass in neutrinos than in baseballs, automobiles, planets, and stars put together? Could our visible world really be only a small subset of the physical reality that is our Universe? Are we just a pocket of light in an otherwise hidden Universe?

• • •

As with finding invisible planets, the key to detecting other varieties of invisible matter lies with gravity. Unfortunately, despite

2. Of course, without the electromagnetic force holding them together, all of the atoms making up your body would fall apart before you had the chance to move your hand through a table. But let's ignore this for the moment.

being one of the longest-scrutinized aspects of nature, gravity remains today the least understood of the four known forces of nature. Aristotle, perhaps the first philosopher to develop and advocate the scientific method of experimentally testing competing hypotheses, proposed one of the earliest theories of gravity. He understood that Earth pulled objects toward it, but his theory was largely incorrect. For example, he argued that if you drop a heavy object and a light object from a given height, the heavier of the two will accelerate faster and reach the ground first. Despite the fact that he essentially invented the scientific method, it seems that he did not apply it very rigorously to his own theory of gravity.[3] A simple experiment, easily performed in any high school physics course, can show that his theory is incorrect. Strangely enough, Aristotle's erroneous conclusions remained unchallenged for some two thousand years. It was not until the seventeenth century—the time of Brahe, Kepler, Galileo, and Newton—that these questions would begin to be answered.

In 1600 the German astronomer Johannes Kepler, although hoping to follow a career in the Lutheran Church, reluctantly took a position as the assistant to the extraordinary Danish astronomer Tycho Brahe. Only a year later, Brahe died, leaving Kepler with the large collection of astronomical data he had accumulated throughout his life. Armed with such data, Kepler slowly began to reject many of the standard astronomical descriptions of the day. Of his many notable discoveries, the most important were his three laws of planetary motion, referred to as Kepler's laws, which are still widely used today. Kepler's laws describe

---

3. This was certainly not the only time that Aristotle did not apply the scientific method to his theories of nature. For example, he argued passionately that the mouths of Greek men contained more teeth than those of Greek women. Aristotle never bothered to test this faulty hypothesis.

planetary orbits as ellipses rather than perfect circles. His laws also predict the speed of a planet traveling in a particular orbit, and those predictions agree with the data collected by Brahe and others remarkably well.

At the same time, Galileo Galilei, the greatest mathematician and physicist of his time, was conducting numerous experiments that refuted the Aristotelian theory of gravity, for example showing that falling objects of different weights accelerate at the same rate. Unlike Kepler and Brahe, however, Galileo lived in strongly Catholic territory. As his discoveries became increasingly threatening to Catholic doctrine, he found himself persecuted by the church. For his support of the Copernican theory that Earth revolves around the Sun, he was condemned to be silent and spent the last years of his life under house arrest.

Despite the immense importance of these discoveries, all of the progress made by Galileo, Kepler, and Brahe put together pales beside the advances made by Isaac Newton. Newton developed a complete theory of gravitation that encompassed both Kepler's laws and the experimental conclusions of Galileo. By demonstrating that all objects are pulled toward each other with force proportional to each of their masses, Newton was able to explain the observations of Galileo. Newton was also able to show that Kepler's laws were a consequence of his theory of gravitation, thus demonstrating that the same force held planets in their orbits and pulled objects toward Earth. With this theory, he transformed physics into an experimentally credible science. Only Einstein, with his general theory of relativity, has since made any improvements on Newton's theory of gravitation.

A working theory of gravity enables one to predict the motion of bodies from the masses they are gravitationally bound to, and vice versa. If you look up the planet Jupiter (or any other planet) in an encyclopedia, you will surely find a fairly precise value for its mass. This number does not come from placing Jupiter on a bathroom scale, but rather from studying the orbits of its moons.

The motions of planets in our solar system are relatively easy to study compared to those of more distant objects, which appear to move much more slowly across our sky. Instead of simply following an object's location with a telescope, the technique of spectroscopy can be used to measure the velocity of an object relative to the observer. This is possible because of the Doppler effect of light. If an ambulance, with its sirens on, is traveling toward you, you will notice a sudden change in the siren's pitch, or frequency, as the vehicle passes you (figure 1.1). This is because the relative motion of the siren has just changed from, for example, thirty miles per hour toward you to thirty miles per hour away from you. The frequency of the siren is shifted as the rela-

FIGURE 1.1. The frequency of sound from a moving source will appear higher or lower depending on whether it is moving toward or away from you. This is known as the Doppler effect. The same is true for light, which enables astronomers to determine how fast distant stars, galaxies, and other sources of light are moving relative to Earth.

tive velocity changes. The same is more or less true for light. If a star emits light at a given frequency, the frequency that we observe depends on the velocity of that star relative to us. Using this relationship, we can measure the velocity of an individual star or galaxy.

The twentieth-century astronomer Vera Rubin spent much of her life making these types of measurements (figure 1.2). Beginning with the work that would become her master's thesis in 1950, she studied the motion of numerous individual stars in galaxies by observing light of a specific frequency that is emitted by hydrogen atoms. By studying a large enough sample of stars, these observations could be used to describe the motion of an entire

FIGURE 1.2. Fritz Zwicky (left) and Vera Rubin (right) are two of the pioneers of dark matter. In 1933 Zwicky was the first to suggest that large amounts of dark matter might exist. The later observations of Rubin and others led to the consensus that most of the matter in our Universe is nonluminous.

Credit: Zwicky photo courtesy of Caltech Public Relations; Rubin photo courtesy of Carnegie Institute of Washington.

galaxy, or a group of galaxies. (Ironically, this would be considerably more difficult to carry out using stars in our own galaxy, the Milky Way, because of our internal perspective. It is much easier to make these observations from outside of a given galaxy. Even as recently as 1980, both the distance between the Sun and the center of the Milky Way, and the velocity at which the Sun travels around the galaxy's center, were not well measured.)

Rubin's measurements suggested that groups, or clusters, of galaxies might be rotating around a previously unrecognized central point within each cluster, not just expanding outward as was believed. To explain this motion, a huge quantity of mass would have to be present to keep the galaxies in their orbits. In fact, she found that the quantity of mass needed to explain the motion was greater than the total mass of the stars within the galaxies. The stars were simply not numerous enough to explain her observations, and she concluded that some sort of invisible matter must also be present. When she first presented her results in 1950 to the American Astronomical Society, her conclusions were not taken seriously despite the experimental evidence she had accumulated. Later, her doctoral thesis would also be largely ignored.

Although it was far from being widely accepted, the idea of invisible, or dark, matter was not entirely new when Rubin published her theses. In 1933 the astronomer Fritz Zwicky (figure 1.2) had first suggested the possibility of dark matter. He, like Rubin, had argued that the rotational velocities of galaxy clusters required much more mass than was present in the form of stars. Other occasional discussions on the topic of missing mass occurred as well. These, however, were carried out by more senior and well respected (and male) astronomers. At the time of her master's thesis, Rubin was a twenty-two-year-old woman who found her unorthodox ideas earning her a poor reputation with many of her colleagues.[4]

For decades following her doctoral thesis, Rubin turned her studies to more mainstream and less professionally risky aspects of galactic spectroscopy. Eventually, however, she returned to her studies of dark matter. In 1970, she (along with W. Kent Ford) published an article entitled "Rotation of the Andromeda Nebula from a Spectroscopic Survey of Emission Regions" in the *Astrophysical Journal*. The paper described her detailed observations of the motions of stars in the Andromeda Galaxy, and demonstrated that for the stars to be moving with the velocities they had, there would have to be as much as ten times more mass in the galaxy than was visible (see figure 1.3). To put it another way, roughly 90 percent of the Andromeda Galaxy consists of dark matter. Over the next years, those results were confirmed by other groups of astronomers. By the 1980s, much of the astrophysics community had come to accept the conclusions of Rubin's work. Even her master's thesis had been vindicated.

• • •

Despite their resistance, astronomers found themselves led to the conclusion that a large fraction of the Universe's mass consists of unidentified, invisible matter. Many astronomers turned to new

---

4. This was not the only time Rubin is said to have experienced difficulties as a woman in the predominantly male world of astrophysics. Later in her career, for example, she was invited by the famous physicist and cosmologist George Gamow to visit him at the Applied Physics Laboratory at Johns Hopkins University. When she arrived, Gamow and Rubin had to meet in the lobby, as women were not allowed in the offices of the laboratory. In 1965 she would become the first woman to be allowed to observe at the Palomar Observatory in California.

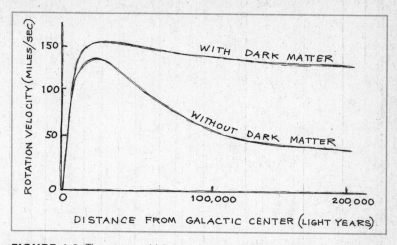

FIGURE 1.3. The rate at which stars in a galaxy revolve around the galaxy's center depends on how much matter is present. Some of the first evidence for dark matter came from the observation that galaxies were rotating more rapidly than they would have been if they contained only stars and other luminous matter.

technologies in an attempt to identify the missing matter. Whereas in the past most astronomy had been conducted using telescopes relying on visible light, by the 1980s methods existed or were being developed to study radio and infrared emissions from stars, as well as x-ray, gamma-ray, ultraviolet, and far-infrared emissions.

These new technologies did not answer the questions raised by Rubin and her colleagues, however. Instead, they only made it clearer that the majority of our Universe was, to us, a mystery. Whether the dark matter consisted of planets, dead stars, black holes, exotic particles, baseballs, Cadillacs, or space monkeys was entirely unknown. The nature of this mystery became the quest of a scientific generation.

CHAPTER 2

DEAD STARS, BLACK HOLES, PLANETS, AND BASEBALLS

Black holes are when God divides by zero.
—Stephen Wright

Convinced that faint or invisible matter makes up the bulk of our Universe's mass, astronomers began to seriously ask what this mysterious matter could be. Although little was known about the nature of dark matter, new and more sophisticated studies of the rotations and motions of galaxies provided astronomers with a more detailed description of its distribution in the Universe. In many galaxies, including our own, the stars form a fairly flat plane or disk. Dark matter has a different distribution. The studies found that dark matter surrounds most galaxies in roughly spherical clouds, called halos. Dark matter halos are significantly larger than the visible part of most galaxies, and often extend well into intergalactic space.

Galaxies are essentially large collections of stars and dark matter held together by mutual gravitational attraction. Because stars were known to comprise most of the visible matter in galaxies, it was natural that many of the first popular candidates for dark matter were types of stars. In order to escape detection, however, they could not be ordinary stars. Instead, the strange dark stars would have to be dead or burned-out stars, or perhaps stars that never even began to burn.[1]

A star that gives off no light (or very little light) is a strange star indeed. To understand such unfamiliar objects, we must first understand the process of stellar evolution (figure 2.1). Stars are not static objects. Rather, they change and develop over time. At each stage of a star's evolution, there is a balance—or in some cases an imbalance—between the star's own gravity, which compresses it, and the force created through nuclear fusion or other processes, which pushes the star outward and apart. Whichever of these forces ultimately prevails determines how that star changes and evolves.

Stars originate in giant clouds of gas. If dense enough, a cloud of gas will collapse under the pull of gravity, forming a newborn star, also known as a protostar. Our Sun formed in this way about five billion years ago. If a protostar is small, with a mass less than about one-tenth of our Sun's mass, the material never gets hot enough to begin burning nuclear fuel, and a dim and unspectacular object called a brown dwarf is formed. A brown dwarf is essentially just a very large planet with a mass between five and ninety times that of Jupiter, made mostly of hydrogen and helium. Brown dwarfs glow only faintly, making them very difficult to observe and good candidates for dark matter.

A denser gas cloud can collapse into a more massive protostar

---

1. Stars, of course, do not "burn" in the same way that wood in a fireplace does, but rather undergo a process called nuclear fusion. Nevertheless, this process is often described as burning.

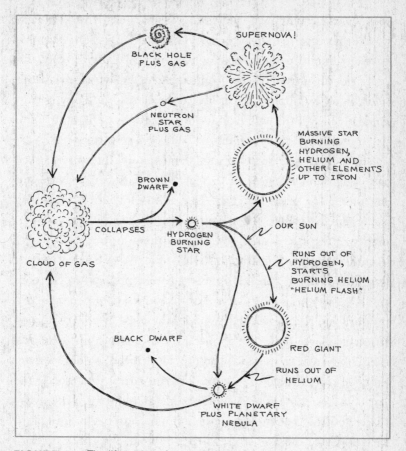

FIGURE 2.1. The life cycles of stars, from their beginning as enormous clouds of gas to various possible endings such as a neutron star, black hole, or dwarf star.

whose core can become sufficiently hot for it to begin burning through nuclear fusion. Nuclear fusion is the process by which the nuclei of lighter elements combine to form nuclei of heavier elements; the most common example in stars is the fusion of hydrogen nuclei to form helium. When this occurs, energy is released. The amount of energy released is tremendous, and is what makes the fearsome hydrogen bomb so powerful. (If we could extract the energy in a more peaceable and contained fashion, we would have an almost limitless supply of clean energy. Unfortunately, this process occurs naturally only at extremely high temperatures. Considerable efforts have been made to develop technology able to generate energy through cold fusion, although this has proven to be an incredibly difficult task.)

In a hydrogen-burning star, the explosive pressure created through fusion balances the gravitational force pulling the star together, and a stable star can persist, at least for a time. How long a star remains stable, and what happens afterward, depends primarily on its size. A small star, smaller than our Sun, will slowly burn its hydrogen until it runs out of fuel. A somewhat heavier star, like our Sun, behaves somewhat differently. When the core of such a star fuses all the hydrogen it can, the star generates less pressure and begins to collapse under the force of gravity. Gravitational collapse causes the temperature of the star to increase, and more of its hydrogen, which had previously been too distant from the core and too cold to fuse into helium, begins to do so. As the star compresses, its core becomes hotter and hotter until it eventually becomes hot enough for helium nuclei to begin fusing into carbon. Once the star crosses that threshold, the core's pressure increases dramatically, and the star expands rapidly in what is called a helium flash. Several billion years from now, our Sun will experience its helium flash. When it does, it will expand so much that it will completely engulf the orbits of Mercury, Venus, and Earth. Such a star, called a red giant, once again becomes stable, with the force of gravity balanced by the pressure generated through the burning of helium and other elements.

A red giant star has an outer layer of hydrogen, but rather than fusing, the hydrogen gradually gets thrown off the star by powerful solar winds. The process populates space with new

clouds of hydrogen gas, enabling new stars to form someday—
nature's way of recycling. The enormous cloud of gas that surrounds the remaining stellar core is called a planetary nebula.
Many of the famous Hubble space telescope images are of these
objects. The Ring Nebula, for example, is a small stellar core surrounded by clouds of colorful gas (figure 2.2).

The collapse of the core of a red giant star recommences when
its helium supply becomes sufficiently depleted, and eventually
results in a dense core of matter—elements such as carbon and
oxygen—that is no longer hot enough to undergo fusion. Although the core has roughly the same mass as our Sun, it has a
volume roughly equal to that of Earth. The density of such a star,

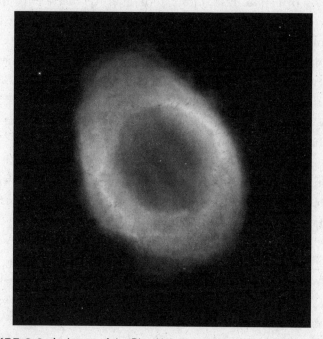

FIGURE 2.2. An image of the Ring Nebula taken by the Hubble Space
Telescope. The Ring Nebula is an example of a planetary nebula.

called a white dwarf, is hundreds of thousands of times greater than that of iron, and its temperature is generally around 100,000 degrees Fahrenheit. By comparison, the Sun's surface temperature is only about 10,000 degrees. With such a high temperature, a white dwarf glows like white-hot coals, even though it is no longer burning any nuclear fuel. Also like hot coals, white dwarf stars slowly cool and become fainter, eventually becoming black dwarfs. Dwarf stars were once thought to be among the most promising candidates for dark matter.

Not all stars become dwarfs, however. Very large stars, considerably more massive than our Sun, are hot enough to burn helium and other elements without becoming red giants. Inside such a star, helium fuses into carbon and eventually into silicon and iron. When all of the stellar matter is fused into iron, the process of fusion comes to an abrupt halt. When this happens, the star can no longer generate the pressure needed to counteract its gravity and the entire star undergoes total gravitational collapse. In a fraction of a second, the star first falls into itself, creating new types of ultra-compressed matter, and then rebounds in an incredible explosion: a supernova.

Supernovae are rare, fortunately. They release so much energy that if one were to occur at a distance less than ten or twenty light years from Earth, the effects on life here could be disastrous. We have historical records of three supernovae having been observed within our galaxy (but sufficiently distant to pose no danger to Earthlings), the most recent occurring in 1604. One that took place nearly one thousand years ago left behind a remnant in the form of the famous Crab Nebula. The Crab supernova was so bright that it could be seen easily during the day, for more than three weeks.

Although some of the star gets blown out into space during a supernova explosion, much of the stellar material remains intact.

That remaining material becomes either a neutron star or a black hole, depending on its mass. A neutron star is an incredibly compressed object with a mass greater than our Sun's, but less than twenty miles wide. Ordinary matter could never be compressed to such a density. The protons in atoms are positively charged and therefore repel each other as they get close together. This repulsion prevents ordinary matter from such extreme compression due to gravity or other forces. Earth's gravity does not compress it to a smaller size because of this resistance. Neutron stars, however, are so massive that gravity can force the protons and electrons in the star together, destroying each other, leaving only neutrons behind. Because neutrons have no electric charge, they can be compressed to ridiculous densities. A chunk of a neutron star the size of a small garbage dumpster would weigh more than a block of iron the size of 100,000 Empire State Buildings.

Neutron stars are not only far denser than typical stars, they also spin much more quickly. Whereas our Sun rotates about thirteen times in a year, a typical neutron star can rotate hundreds or thousands of times every second. Neutron stars spin so vigorously for the same reason that a figure skater spins faster as she pulls her arms in toward her body (figure 2.3). A basic law of physics is that in the absence of some external force, the angular momentum of anything, whether star or skater, must remain constant. Angular momentum is essentially the product of the distance of a mass from an axis and the velocity with which it spins. So when a skater pulls her arms toward her body—that is, the distance of her mass from her center decreases—the conservation law predicts that she will spin faster. (You can test this while skating, or by spinning in an office chair while alternately extending and retracting your legs.) Similarly, as a star collapses and becomes more concentrated, it spins much faster than it had before.

If the star is heavier yet, even this all-neutron matter will not

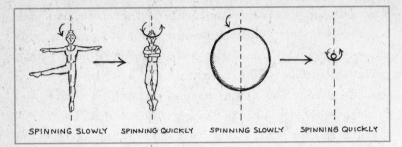

SPINNING SLOWLY   SPINNING QUICKLY   SPINNING SLOWLY   SPINNING QUICKLY

FIGURE 2.3. Just as a figure skater spins faster when she pulls her arms and legs in toward her body, a star spins more rapidly after it collapses. A neutron star formed in such a collapse can rotate hundreds or thousands of times per second.

be able to resist the force of its own gravity, and it will collapse further, becoming a bizarre object known as a black hole. The concept of black holes has a long history, extending back to at least the eighteenth century, when an English geologist named John Mitchell showed that if an object was sufficiently massive and dense, gravity would become so strong that even light could not escape from it. Mitchell named this kind of speculative object a "dark star."[2]

Although Mitchell's dark stars were indeed an intriguing idea, he didn't have the tools needed to fully develop it. The eighteenth-century understanding of gravity (Newton's theory) was unable to describe the conditions under which Mitchell's dark stars could exist. Such an object is just too dense, and its gravity too powerful, for Newton's theory to be applied. The Newtonian theory of gravitation, although very successful at describing the orbits of

---

2. Mitchell's idea gained enough attention to merit inclusion by Pierre Simon Laplace, a prominent French mathematician, in his astronomy guide of 1795. But speculations didn't remain current for long, disappearing from Laplace's book in its third edition.

planets and the motions of bodies in our everyday experiences, has its limitations. The scientific revolution that was Einstein's theory of relativity would be needed to grasp the inner workings of a black hole.

• • •

In 1915 Albert Einstein published his general theory of relativity. With it, he showed that gravity was not simply the force that pulled masses together, as Newton had thought. Rather, gravity is a consequence of the geometry of space and time, or space-time.

Einstein's space-time, a concept he had put forth in his earlier special theory of relativity, was the combination of the three dimensions of space (north-south, east-west, and up-down, so to speak) with a fourth dimension, time. The special theory of relativity showed that what appeared to one observer to be an object moving through space could be perceived as motion through time, or some combination of time and space, to an observer watching from a different vantage point. Different observers might measure different distances separating the same two points in space. Furthermore, neither observer would be wrong. Einstein showed that the distance between two places depends on your frame of reference. Similarly, the length of time that passes between the same two events can be different to observers in different frames of reference.

With these notions of the relativity of distances and times, it might seem that science had lost its objectivity. If two people would not agree about the distance between, say, New York and Paris, how could we decide what value to put in an atlas, teach to schoolchildren, or use to calculate the best route for an airplane to take? Practically speaking, however, this is not a major problem.

All observers on Earth that measure the distance between New York and Paris will very nearly agree, with the effects of relativity being vanishingly small. Only those observers traveling at speeds close to the speed of light will see any significant deviation in their measurements of Earth's geography. But what about those very-high-speed observers? Can they measure different distances between the same two points and both be correct? Does relativity prevent physics from giving us objective answers? Does relativity say there is ultimately no "truth" in nature?

No. The apparent problem is really only skin deep. Einstein showed that if space and time are not considered as separate entities to be treated independently, but as a single concept of space-time, objectivity is restored. Distances in space-time are agreed upon by all observers in all frames of reference, even if distances in space and lengths of time are not.

Under the laws of Newtonian physics, objects that are free from the influence of forces such as gravity always move along straight lines. Another way of saying this is that the shortest path between any two points along which an object travels will trace the trajectory of the object. Einstein's new theory, in contrast, said that this is not strictly true in space, but rather in space-time. When gravity or other forces are not present, the shortest possible path in space-time, called a geodesic, is identical to the straight line predicted by Newton's theory. When large quantities of mass or energy are present, however, the geometry of space-time becomes curved, or warped, and the geodesics that objects travel along become modified. The curvature, which causes the trajectories of objects change in the presence of a mass or energy, is what we know as the force of gravity.

To understand what I mean by the curving or warping of space-time and how it generates the effect we call gravity, consider the following analogy. A taxi that charges the same fare for each minute driven is asked to take a customer from the airport to a hotel across town. Assuming that the taxi driver has the passenger's best interests in mind, and that the taxi can drive at the same speed on every street, she takes a route that is as close to a straight line as possible between the airport and the hotel. This is analogous to traveling through flat, or uncurved, space-time.

But now imagine that the taxi cannot drive the same speed everywhere in town, but instead varying amounts of traffic slow down travel along some routes. Again assuming that the taxi driver has her passenger's interests in mind, she will take a route that goes around the areas of highest traffic. The fastest, and cheapest, possible route is the taxi's geodesic.

Taking this analogy further, let's imagine that a single central point in the city has the very worst traffic, with the amount of traffic steadily decreasing as one goes away from the center. In this case, it would be possible that the fastest route for a taxi to take across town would be an arc resembling a half circle. If, after arriving at the hotel, the driver returns to the airport to pick up another passenger by traveling in the other direction around the center, the taxi will have made an orbit around the city center.

In this taxi picture, the distances between two points in the city are just like distances in general relativity. If there is no traffic, that is, if there is no gravity, then taxis and other objects move in straight lines through space. But if the taxi encounters different patterns of traffic on two routes from the airport to a hotel, then the taxi charges different amounts of money to cross the same distance (because it travels faster or slower depending on the traffic). The difference in cost for traversing the same linear distance is analogous to saying that even if two distances of space are equal, gravity can cause them to be of different distances in space-time. And because, according to Einstein, it is the shortest distance in space-time rather than space alone that an object's trajectory fol-

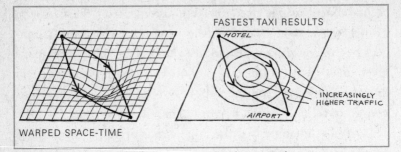

FIGURE 2.4. Much as heavy central traffic congestion may make the "shortest" route for a taxi a curved trajectory (above, right), the presence of mass curves or warps space-time according to Einstein's general theory of relativity (above, left). This causes objects to move in arcs or in orbits around massive objects, instead of in straight lines through space. For example, the curvature of space-time causes planets to orbit the Sun.

lows, a planet orbits around a star, just like the taxi takes a route around the city's center (figure 2.4).

Earth orbits the Sun because of the Sun's gravity. If not for gravity, Earth would move along a straight line. The Sun's gravitational field has warped the surrounding space-time, however, resulting in a geodesic that is not that straight line, but rather is an ellipse that curves around it. That is the path of Earth's orbit.

In 1916, shortly after Einstein's publication of his general theory of relativity, the idea of dark stars was revitalized by the German astrophysicist Karl Schwarzschild. Using Einstein's equations, Schwarzschild found that if a mass were concentrated enough, gravity would make it not merely difficult for objects to escape its pull, but impossible. Using the taxi analogy once again, it would be as if the city's center had a traffic jam so bad that it would take an infinite amount of time (and an infinite amount of the passenger's money) to drive away from the center.

According to Schwarzschild's solution to Einstein's theory, a sufficiently dense object acts as if it has a spherical shell through which no matter or light can ever escape, just like a taxi can never navigate the city's horrendous traffic jam. A circle drawn on a map surrounding the extent of the jam is analogous to the shell, which became known as the Schwarzschild radius or event horizon. Anything that falls within the Schwarzschild radius can never be retrieved. The gravity is so strong that an inescapable point of no return is created. The term "black hole" was later coined by the cosmologist John Wheeler to describe these incredible objects.

To appreciate the weirdness of a black hole, imagine falling into one. The force of gravity would continue to accelerate you as you fall though the Schwarzschild radius and into the center of the black hole (of course you would have been killed by the incredibly strong force of gravity, but let's ignore that detail). An outside observer, however, would see something very different. Seen from a stationary perspective outside of the black hole, you would appear to begin falling toward the object but would slow down as you approached its Schwarzschild radius. No matter how long an observer watched, he would never see you pass through the Schwarzschild radius and into the black hole. According to Einstein's theory, the way that time passes to two observers can be different. In the presence of a black hole, one observer (the one falling in) sees himself falling in quickly, while the other observer (the stationary one outside) sees time stretched in such a way that the falling person never passes through the Schwarzschild radius and into the black hole.

As a person approaches the Schwarzschild radius of a black hole, the light being reflected off his body loses more and more energy as it struggles to escape the black hole's enormous gravity. As the person gets closer and closer to this radius, and the escaping light becomes less energetic, the image of the person falling toward the black hole slowly fades from the point of view of the outside observer, eventually becoming invisible.

At the time that Schwarzschild presented his black hole solution of Einstein's equations, it was only that—a solution to equations. The existence of the solution meant that black holes were possible, but it certainly did not mean that they existed. To create a black hole, an enormous quantity of mass has to be forced into a tiny volume. For example, to compress Earth enough for it to become a black hole, you would have to fit it into a sphere with a radius of about one centimeter. A star with the mass of our Sun would have to be crushed into an object a few kilometers across to become a black hole. Because of the obvious difficulty of doing something like that, even after Einstein's equations were shown to allow for the possibility of black holes, most astrophysicists thought it unlikely that black holes actually existed. Many argued that such densities could never be achieved in nature, or that Einstein's equations would not apply under such extreme conditions and that a new theory, without such wild predictions, would be needed to describe these circumstances.

In 1974, when it remained unknown whether black holes existed, the theoretical physicist Stephen Hawking made a bet with astrophysicist Kip Thorne that the star Cygnus X-1 was not a black hole (figure 2.5). Eventually, it became clear not only that black holes existed, but that the star Cygnus X-1 was one. Hawking conceded his bet in 1990, and Thorne has since received his prize, which was a year's subscription to *Penthouse* magazine. Hawking didn't learn his lesson, however, making a new bet in 1991 with Thorne and John Preskill that naked, or observable, singularities, the strange phenomena thought to be at the heart of black holes, did not exist. Again, Hawking conceded his bet in 1997, in typical Hawking style, "on a technicality."

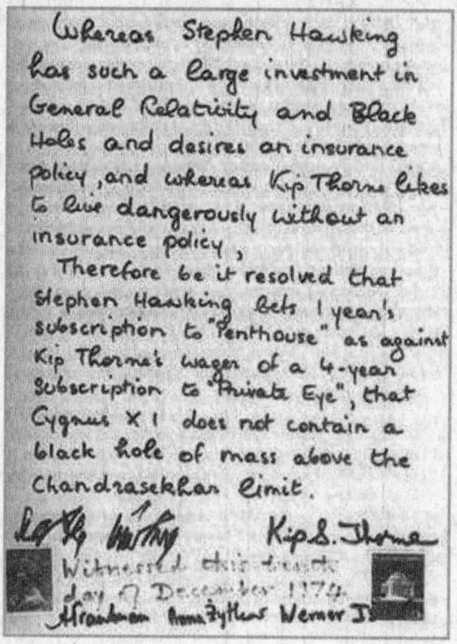

Whereas Stephen Hawking has such a large investment in General Relativity and Black Holes and desires an insurance policy, and whereas Kip Thorne likes to live dangerously without an insurance policy,

Therefore be it resolved that Stephen Hawking bets 1 year's subscription to "Penthouse" as against Kip Thorne's wager of a 4-year subscription to "Private Eye", that Cygnus X 1 does not contain a black hole of mass above the Chandrasekhar limit.

Kip S. Thorne

Witnessed this [illegible] day of December 1974.

FIGURE 2.5. The wager made between Stephen Hawking and Kip Thorne over whether Cygnus X-1 was or was not a black hole. Hawking has since conceded the bet and provided Thorne with his prize.

Credit: Courtesy of Kip Thorne.

Today, there is a wide array of evidence for the existence of black holes. Many of these black holes were most likely produced in the supernova explosions of very massive stars. Additionally, much larger black holes are known to reside in the cores of galaxies. At the center of our own galaxy, for instance, there is a colossal black hole, more massive than two million Suns. Such black holes are thought to exist in the centers of most galaxies, and were most likely formed through a series of mergers between smaller black holes, stars, and other matter over the course of time.

• • •

Not all candidates for dark matter need be objects that result from the process of stellar evolution. Astronomers believe that early in the history of the Universe, small black holes may have formed directly without having been stars first. Such primordial black holes could remain scattered across the Universe today. If they formed in large enough quantities, they could be the dark matter's primary component.

Around 14 billion years ago, shortly after the Big Bang, the Universe was extremely hot and extremely dense—much more so than even the cores of stars—and was made up not of atoms or even protons and neutrons, but rather of their building blocks: quarks, gluons, and other particles, mixed in a physical state known as plasma. Eventually, as the Universe expanded and cooled, the quarks and gluons joined together to become the ordinary stuff that makes up the atoms in our present world. Random fluctuations in the temperature and density of the plasma could result in particularly dense pockets in space. Those pockets, if concentrated enough, could have attracted large enough quantities of matter to form black holes—albeit ones much smaller than those that would be formed through stellar evolution. In fact, these primordial black holes could be as light as a billionth of a billionth of the mass of the Sun (still about a billion tons). Could those small black holes be what we call dark matter?

Such a conclusion would be bolstered if we knew that such ancient black holes would have survived the 14 billion years from the Big Bang to the present day. Up until the 1970s, it was thought that a black hole, once formed, could never be destroyed or even reduced in mass. According to Einstein's theory alone, any material that falls inside of the Schwarzschild radius of a black hole could never escape. Einstein's theory alone does not account for the strange effects of quantum physics, however.[3] In 1974, by

---

3. We will return to the subject of quantum physics in the next chapter.

studying these strange effects, Stephen Hawking showed that is was not impossible for mass to escape from a black hole.

In his most important scientific work, Hawking argued that the laws of quantum physics allowed for pairs of particles to appear spontaneously near, but within, the Schwarzschild radius of a black hole. One would be a particle of matter, such as an electron. The other would be a particle of antimatter, such as the positive version of the electron, called a positron. When such particles pop into existence in most parts of the Universe, the matter and antimatter particles quickly annihilate each other and thus do not violate the law of conservation of energy for more than a fraction of a second. When particles appear at the edge of a black hole, however, it is possible for one of the two to travel away from the black hole, eventually escaping, while the other is pulled back in. The two particles never come into contact with each other, and therefore are never able to annihilate each other. Every time a particle escapes in this way, the black hole loses a minuscule amount of its mass, because one can think of the particle as having been emitted by the black hole. Over a long time, this process, called Hawking radiation, could potentially shrink a primordial black hole and eventually cause it to evaporate completely.

The rate at which a black hole evaporates depends on how big the black hole is. A large black hole, like the kind that might be left behind after a supernova explosion, undergoes the process of Hawking radiation at a very slow rate, and essentially retains all of its mass. Smaller black holes radiate more quickly, and as they become smaller, they shed particles at an ever-quickening rate. In the last moments of their life, black holes radiate away enormous amounts of energy in the form of high-energy particles. Astronomers watch for signs of these particles, such as gamma rays and neutrinos, in the hopes of collecting evidence for the existence of primordial black holes. Although no such evidence has been found so far, many astronomers are still searching for primordial black holes. Today, we simply do not know whether these objects exist in the Universe, or if they do, how much of the dark matter they might constitute.

Astronomers call the dark matter candidates I've described in this chapter massive compact halo objects, or MACHOs. The group includes dead stars like white dwarfs, stars that never burned like brown dwarfs, strange exotic stellar entities like neutron stars and black holes, and large Jupiter-like planets. To be complete, however, the classification of MACHOs can include any piece of concentrated matter that resides in galactic dark matter halos. Some astronomers are fond of including in their lists of possible MACHOs such things as Cadillacs, thesauruses, baseballs, and other everyday objects, although they are certainly unlikely to have significant abundances throughout the Universe.

Searching for MACHOs has been a pursuit of observational astronomers for decades. If enough MACHOs could be found, perhaps the identity of the dark matter could be determined once and for all. Although today we have incredibly powerful telescopes like the Hubble Space Telescope that have the power to see some types of nearby MACHOs, the chances are that you won't find any if you don't know where to look. To know where to point a powerful telescope like the Hubble, astronomers take advantage of a relativistic phenomenon known as gravitational lensing.

Without Einstein's theory of general relativity, the idea of a gravitational lens is an impossibility. According to Newtonian physics, the strength by which the force of gravity pulls two bodies together is proportional to the masses of the two bodies—the greater the mass, the stronger the gravitational force. Light is known to have no mass (and I'll discuss how we know that in chapter 5); as a result, Newton's theory predicts that light would not experience gravity. Einstein's theory of relativity makes a different prediction: gravity influences the motion of anything with energy, even if it does not have any mass. Thus the trajectory of light is altered when it passes by heavy objects, such as stars.

When Einstein's general theory of relativity was first published, the only very massive objects that were known were also very bright, such as the Sun. The Sun's brightness, however, makes it next to impossible to see any light rays being bent as they pass close by. To test Einstein's theory, a very massive, nearby dim object was needed. Fortunately, on May 29, 1919, exactly such an object was in the sky, for on that day a total solar eclipse provided an opportunity to use the Sun as a gravitational deflector of light, with its own brightness largely concealed by the moon.

The eclipse of 1919 was particularly well suited to test Einstein's theory. For one thing, it lasted longer than most eclipses—about six minutes. For another, it took place in almost perfect alignment with a cluster of stars called the Hyades that provided an excellent source of light to be bent by the Sun's mass. Arthur Eddington, Britain's leading astronomer at the time, led an expedition to an island off the coast of western Africa to observe the eclipse and thereby resolve whether Einstein's or Newton's theory of gravity was most correct.[4]

Eddington's results were announced in November 1919 at a special meeting of the Royal Astronomical Society in London. The light from the Hyades had indeed been bent by the Sun's

---

4. Eddington's expedition very nearly did not take place. In 1917, Britain reintroduced mandatory conscription into the military. Being a pacifist, Eddington refused, and for this faced the possibility of serious punishment. Only after the intervention of Sir Frank Watson Dyson, the Astronomer Royal at the time, was Eddington given the freedom to travel and carry out his experiment.

gravity. Einstein, his theory confirmed, instantly became world famous.

Eddington's technique is quite similar to the gravitational lensing technique used by modern astronomers to search for dark matter in our galactic halo and in other nearby galaxies. If a bright source of light is nearly perfectly aligned with a MACHO along our line of sight, the source's light will deflect slightly when it passes the MACHO, curving around the object and potentially toward us. The MACHO acts as a lens, focusing the light from a distant bright object onto a point, which makes the light source temporarily much brighter and more easily seen by astronomers (figure 2.6). Once such a lensing event is detected, astronomers look more carefully at that point in the sky, in hopes of finding the MACHO that is responsible. By searching for events of gravitational lensing, astronomers can look for MACHOs over much larger portions of the sky than could be done with traditional telescope searches.

In 1993 a collaboration of scientists began the largest and most detailed search for stars being focused by MACHOs acting like a gravitational lens. The astronomers, operating a telescope with a fifty-inch diameter at the Mount Stromlo Observatory in Australia, constantly monitored millions of stars in our galaxy and in

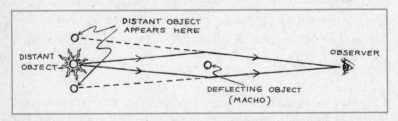

FIGURE 2.6. Light from a distant object can be deflected gravitationally by massive objects along its trajectory. This phenomenon, called gravitational lensing, can be used to search for dim or invisible objects that might make up dark matter, such as dwarf stars, neutron stars, or black holes.

nearby galaxies, looking for individual stars that would become much brighter for a time as a result of being gravitationally lensed. Although such events are incredibly rare, when they do occur a star is often dramatically brightened for days, weeks, or even months.

The Mount Stromlo telescope recorded data from many possible gravitational lensing events over eight years of observation. The astronomers found that most of these events appeared to have been caused by white dwarf stars or similarly heavy MACHOs, although some may also have been generated by smaller objects, such as brown dwarfs or Jupiter-like planets. Although dozens of these events were recorded, lensing observations alone could not conclude with certainty what kind of objects were responsible for the brightening. An estimation of the object's mass could be inferred, but little more. To identify the source of the gravitational deflection, another technique was needed. To identify a potential MACHO, astronomers would follow up lensing observations with the Hubble Space Telescope to reveal exactly what was bending the light. The first conclusive identification of a light-bending MACHO came in December 2001, when Hubble revealed a small white dwarf star about six hundred light years away—practically in our backyard by astronomical standards. It was thus certain that at least some fraction of the Universe's dark matter consists of dwarf stars.

• • •

As more gravitational lensing data accumulated, the astronomers of the Mount Stromlo group argued that although they found few lighter objects, like planets or brown dwarfs, enough white dwarfs and similar objects were found to possibly constitute as much as 50 percent of our galaxy's dark matter. Their evidence left many astronomers unconvinced. First, all these white dwarf stars would once have been ordinary stars that eventually ran out of fuel. Over their lifetimes, such stars would have given off substantial amounts of gas, heavy chemical elements, and infrared radiation, none of which has ever been seen. Furthermore, there are also ways in which observations might resemble gravitational lensing events without a MACHO being responsible. Such occurrences are expected to be rare, but they might be frequent enough to make the number of MACHOs in our galaxy considerably smaller than the estimate of the Mount Stromlo group. A greater number of clearly identified MACHOs, such as the dwarf star seen by the Hubble telescope, would be needed to reach a firmer conclusion. Tragically, in January of 2003 bushfires ravaged parts of Australia, resulting in the destruction of more than 500 homes, the deaths of several people, and the devastation of much of the Mount Stromlo observatory.

• • •

Most types of MACHOs face one other major shortcoming, called the nucleosynthesis problem. Recall that early in the history of the Universe, all of space was filled with a quark-gluon plasma, which cooled to form protons and neutrons. As the temperature dropped further, some of these protons and neutrons became bound together as well, forming the nuclei of the lightest chemical elements in the periodic table. This process, called nucleosynthesis, was proposed by Ralph Alpher, Hans Bethe, and George Gamow in their famous "Alpher-Bethe-Gamow" paper of 1948. (Apparently, the eccentric Gamow added Hans Bethe's name to the list of authors in order to resemble the first three letters of the Greek alphabet.) In the paper, often called the "alpha-beta-gamma" paper, the authors showed that the relative abundances of the light elements, such as hydrogen, lithium, helium, and deuterium (which is hydrogen with a neutron in its nucleus), could be reliably calculated within the context of the Big Bang theory. At the time of their study, the Big Bang theory was still controversial. When the predictions of the Alpher-Bethe-Gamow paper were found to match the observations of the quantities of light elements in our Universe, it elevated the status of the Big Bang to that of a much more respectable, and experimentally verified, theory.

The results of Gamow and his collaborators present a problem for MACHOs as dark matter, however. Their calculation can be used to show that the total amount of matter in the Universe made up of protons and neutrons (including all atoms and molecules) must be considerably less than the amount of dark matter known today to exist. There could not possibly be enough white dwarfs, neutron stars, planets, and other such stuff to make up all of the dark matter. Whatever the dark matter consists of, the findings of Gamow and his collaborators show that it cannot be made up of protons, neutrons, atoms, or molecules.[5]

All of this may leave you a bit disappointed. In this light, dwarf stars, planets, neutron stars, and black holes seem to be rather poor candidates for the dark matter of our Universe, to say the least. But if MACHOs are not likely to comprise dark matter, what else could it be?

---

5. An exception to this rule is found in the case of primordial black holes. Such objects would have formed prior to the generation of the light elements, and therefore are not limited to the quantity of protons, neutrons, atoms, and molecules present after nucleosynthesis took place.

CHAPTER 3

DARKNESS FROM THE QUANTUM WORLD

We are all agreed that your theory is crazy. The question which divides us is whether it is crazy enough.

—Niels Bohr

The demise of the MACHO hypothesis does not leave us without good candidates for dark matter. There are other possibilities that do not include any chunks of rock, burnt-out stars, or even any black holes. Rather than among the very large, dark matter candidates can be found among the minute members of the particle zoo that has emerged during the last century from the strange world of the very small—the world of quantum physics. Quantum physics is going to dominate many of the chapters throughout this book, so before I discuss which types of quantum particles might make up the dark matter of our Universe, I'm first going to give a short tutorial in some of the ideas of quantum theory. If you find yourself confounded as what begins as a strange theory gets stranger and stranger, try to remember that many physicists have often felt the same way.

The discovery of quantum physics required physicists to question much of what they thought they knew about the world. Their intuition about how things are, or should be, had to be thrown aside as they reconsidered the work of every physicist from the time of Newton to the close of the nineteenth century. The need for a new theory is, in hindsight, obvious because it is now clear that by the early twentieth century, classical physics was incapable of explaining theoretically much of what physicists were seeing empirically. This was not always so clear, however. Not surprisingly, along with quantum theory's revolutionary description of nature would come one of the most contentious debates science has ever seen. However obvious the need for a new theory might seem today, most physicists at the end of the nineteenth century felt that the laws governing nature were quite well understood. For example, in the year 1900 the prominent British physicist Lord Kelvin claimed in an address to the British Association for the Advancement of Science that "there is nothing new to be discovered in physics now. All that remains is more and more precise measurement."[1]

He couldn't have been more wrong. Ironically, in the quantum world in which physicists were about to find themselves, precise measurement is not always a possibility. For the odd creatures of the quantum zoo, quantities such as energy, time, position, and velocity often cannot be well defined: a specific object need not possess a specific quantity of energy or exist at a specific location.

---

1. This would not be the only time Lord Kelvin's scientific judgment would prove to be wildly flawed. He strongly opposed Charles Darwin's theory of evolution and natural selection, preferring to remain "on the side of angels." He also confidently declared, "I can state flatly that heavier-than-air flying machines are impossible."

Instead, quantum particles can have multiple, and sometimes very different, energies, locations, and velocities—and all at the same time.

The first hints of the need for a new physics came from the study of a class of objects we call blackbodies. A blackbody is a perfect absorber and emitter of energy; that is, a blackbody gives off all of the energy it takes in. The Sun is a nearly perfect blackbody, as is a hot coal. Nineteenth-century physicists thought that it should be possible to precisely calculate the spectrum of light radiated from a blackbody using only the most basic principles of classical—that is, Newtonian—physics. The only problem was that blackbodies did not emit energy in the way that classical physicists predicted. According to their model, blackbodies should emit high-frequency light, which they did not actually emit. Nevertheless, most physicists remained confident in the foundations of their science, and were reluctant to consider any radical solution to the problem of the blackbody spectrum.

But years passed and no conventional solution to the problem was found. In the 1890s Max Planck, a German physicist, worked on the problem of blackbody radiation and even produced a formula that matched the blackbody observations. It was not a very compelling solution, however, because it was not based on any fundamental principles. Even Planck himself considered this accomplishment to be only "lucky intuition," with no real underlying understanding of the problem. What Planck and others wanted was not only to be able to describe the behavior of these bodies, but to understand why they acted the way they did.

Planck worked on this problem for years, and in 1900 succeeded. To accomplish this, however, Planck made a rather strange assumption about the physics of blackbodies and light that had been implicit in his mathematics. This assumption would go down in the history of science as perhaps the first moment of the quantum

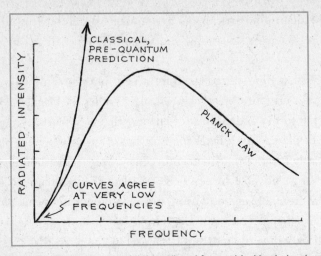

FIGURE 3.1. The spectrum of light radiated from a blackbody (such as the Sun) as compared to the spectrum predicted by classical physics. This disagreement was one of the first observations to eventually lead to the development of quantum theory.

revolution. He showed that the correct blackbody spectrum could be predicted from basic principles alone if one assumed that the matter making up a blackbody could not absorb or emit energy in any arbitrary quantity, as it could according to the laws of classical physics, but only in specific and discrete amounts. In his calculation, individual "pieces" of light carried a fixed quantity of energy proportional to their frequency (figure 3.1).[2]

The revolutionary nature of Planck's assumption might be why a further implication of his hypothesis was overlooked at the time: that light possesses not only the properties of a wave but also the properties of an individual particle. Describing light as a particle strongly conflicted with the theory of light current at the

---

2. To see how the blackbody spectrum calculated by Planck changes with the temperature of the blackbody, take a look at http://webphysics.davidson.edu/alumni/MiLee/java/bb_mjl.htm.

time, in which light was treated as a wave. The conflict, however, was generally ignored. Even Planck himself was reluctant to adopt a new and unorthodox view of the nature of light and continued to think of his calculation as merely a mathematical trick.

Planck's hypothesis and his calculation of the blackbody spectrum were, nevertheless, a crucial step from the classical description of nature toward a new quantum theory. That said, if I had to pick an exact date for the eruption of the quantum revolution, I would not choose the date associated with Planck's discovery. Instead, I would say that it began in 1905 with Albert Einstein's interpretation of the photoelectric effect. With that new and radical idea, Einstein, as he did a number of times in his career, turned physics on its head.

In 1905, Einstein, having failed to find employment as a teacher of physics or mathematics, had been working for a few years as a patent clerk in Bern, Switzerland. Having wanted an academic appointment, he undoubtedly found his job disappointing, but the seeming inopportuneness allowed him ample time to focus his thoughts on his physics. The extra time and freedom paid off when in a single issue of *Annalen der Physik*, a German physics journal, he published three seminal articles. One introduced his special theory of relativity. One proved that atoms existed. And another described his interpretation of the photoelectric effect.

The photoelectric effect is familiar to anyone who has used a solar-powered calculator. Light, striking the proper material connected to an electrical circuit, induces an electric current. According to classical physics, if the intensity of the light were increased, more electrons should be knocked free and a larger electric current would be generated around the circuit. The results of an experiment in 1887, however, did not match the expectations of classical physics. Whether an electric current was generated around the circuit depended not on the light's intensity but on its frequency. Incoming light would induce an electrical current only if its frequency was greater than a certain amount; increasing the intensity of a beam increases the electrical current only if it contains light of a sufficiently high frequency.

Einstein had been searching for an explanation for this phenomenon. Intrigued by Planck's work, Einstein suggested that the relationship between the energy and frequency of light identified by Planck was not the result of how a blackbody operated, but rather was an intrinsic property of light itself. He proposed that waves of light were actually made up of individual pieces, which he called quanta (singular, quantum). Because an individual quantum of light with a higher frequency had more energy than one with a lower frequency, only certain light quanta—those with an adequately high frequency—would have enough energy individually to free an electron from an atom, causing the photoelectric effect. No matter how many low-frequency light particles you showered onto a solar array, no electrons would be freed, and no electric current would be generated (figure 3.2).

Einstein's quantum of light, later to be called the photon, was the first example of an object that had the properties of both a wave and a particle. Following Einstein's explanation of the photoelectric effect, many physicists sought to resolve the question of whether light was a wave or a particle. Such a question misses the stunning importance of Planck's and Einstein's work, however. What these revolutionary realizations demonstrated is that light is neither a wave nor a particle, but something new, with both particle-like and wave-like properties: for lack of a better name, it is called a particle-wave. The concept of a particle-wave is one for which we have no intuitive understanding. The features of particle-waves are not obvious in the day-to-day experiences of the macroscopic world. Perhaps unsurprisingly, it would be de-

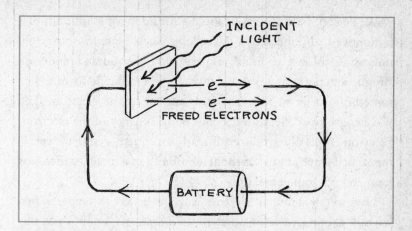

FIGURE 3.2. In the photoelectric effect experiment, light shined onto a metal plate frees electrons and allows electric current to flow around the circuit. Although the predictions of nineteenth-century physics suggested otherwise, the frequency of the light determines whether a current is or is not produced. This was one of the first observations leading to the development of quantum theory.

cades before the particle-wave nature of light would be fully accepted or understood by the scientific community.

Despite the resistance in accepting the particle-wave nature of light, it was not too long before someone found that light was not the only thing that acted like both a particle and a wave. In 1924 a young French graduate student, Louis V. de Broglie, presented his doctoral thesis, in which he argued that all matter, as well as light, demonstrated both particle-like and wave-like properties, and that the heavier a particle was, the more subtle were its wave-like properties. Photons, being very light (please excuse the pun), display very obvious wave-like properties.[3] Tennis balls act as a

---

3. Modern particle physics has determined that quanta of light (photons) must be precisely massless, a point to which I will return in a few chapters. At the time of early quantum physics, however, it could only be said that they were considerably less massive than electrons and other known particles.

wave, but with a wavelength about a millionth of a billionth of a billionth of a billionth of a meter long! Such a length is virtually impossible to observe in any experiment. That doesn't render de Broglie's hypothesis invalid, however, just difficult to test. For something easier to test, physicists turned to the electron. The wavelength of an electron depends on the energy of the electron. Electrons typically produced in early twentieth-century experiments had wavelengths of about a billionth of a meter, which was long enough to be observable.

But to talk about an electron wave seems very strange. When we talk about a wave of water in the ocean, we know what this means: the height of the water varies from place to place, with the high and low points moving along in some direction. Some parts of the wave are higher than others because there are more water molecules at those places. With an electron wave, this picture of things no longer makes any sense. An electron wave is a single object and doesn't consist of parts that are distributed over some region of space like the molecules in a water wave. If you did an experiment to find out where the electron was, you could measure its location very precisely. So, rather than describing how much of something exists at a given point, an electron wave describes the probability of finding the electron at a given point when an observation is made.

The implications of that statement are staggering. For one thing, we can no longer think of an electron as a clearly defined object, at one place at one time. Rather, an electron—and every other quantum object—is more like a wavy cloud of existence, smeared out over space and time. If you know the state of an electron, you can predict the probability of finding it at a given location, or the probability of measuring it to have some specific velocity or other property—but only the probability of a value. You cannot generally determine the precise position, velocity, or other property of a quantum object before it is measured, because that quantum object exists as a sort of combination of different states with different locations, velocities, and other properties. Stranger still, upon measuring the position of such an object precisely, you automatically lose any knowledge of its velocity, and vice versa. This ignorance is not the result of our limited vantage point or technical know-how, but is built into the very nature of every quantum object in the Universe: it is impossible for a particle-wave to have simultaneously a precise position and a precise velocity.

Within the framework of quantum physics, the act of observation itself takes on a new and important role. Until an observation is made, quantum particles simply do not possess the well-defined characteristics that we expect of billiard balls and other macroscopic objects. These bizarre conclusions left many scientists very uncomfortable. To some, physics had undergone a transition from the somewhat unexpected to the completely counterintuitive and seemingly ridiculous.

Of all the strange properties of quantum theory, the one that bothered many physicists the most was the new theory's lack of predictability. Unpredictability, or indeterminism, itself was nothing new. Consider a tossed coin. No practical calculation can tell you in advance whether it will land heads or tails. The best that can be done is to tell you that there is a fifty-fifty chance of getting either side of the coin, because the physical environment is simply too complicated to make a more accurate prediction. To do better, you would need to know the precise location and velocity of every air molecule in the room, and know exactly how the coin was to be flipped and caught. Even if you knew all of this, to perform the calculation you would need a computer far more powerful than anything ever built. Practically speaking, even classical physics can be unpredictable.

Such practical indeterminism, however, does not often raise philosophical objections, because any unpredictability within classical physics is merely the consequence of our ignorance. An omniscient being in the world of classical physics would know of no such thing as probability, only certainty. A hundred years before the discovery of quantum physics, Pierre Simon Laplace, an eminent French physicist, argued that if you knew the location and velocity of every particle in the Universe at a single time, you could in principle calculate the location and velocity of those particles at any time simply by running the equations of classical physics forward or backward.

Quantum theory does not admit such claims. Even if you knew everything about an electron and its environment, it might still be impossible to say where that electron is. All you can possibly predict are the probabilities of finding it at different places. It doesn't matter if you know everything that there is to know about the electron's environment. It doesn't matter if you have an infinitely powerful computer. It doesn't matter if you are an omniscient being. The electron's location cannot be predicted with certainty.

This kind of quantum indeterminism did not sit well with some scientists. Einstein in particular expressed discomfort, asserting that "God does not play dice." Einstein's view of the Universe was that of a perfect and elegant expression of the divine. He was unable, or at least unwilling, to accept that someone knowing these divine laws of nature could not describe the Universe and everything in it completely and with perfect accuracy.

Others, although certainly agreeing that the behavior described by quantum theory was strange, did not find it so philosophically revolting. Niels Bohr, a Danish physicist and quantum theorist, was content to acknowledge the bizarre nature of the new science. Indeed, he expected others to do the same. When faced with an audience of philosophers who, upon first hearing of the new quantum science, remained calm and reacted without surprise or shock, Bohr declared, "Anyone who is not dizzy after his first acquaintance with the quantum of action has not understood a word." Throughout the 1920s and 1930s, Bohr, Einstein, and others would periodically clash over the issues of quantum indeterminacy. In the most famous exchange, Bohr, after listening to Einstein's stubborn refusal to accept indeterminism in nature (or God's playing dice), demanded that his colleague "stop telling God what to do!"

• • •

In his quest to reintroduce determinism into the structure of quantum physics, or to overthrow the theory entirely, Einstein devised clever scenarios, so-called *gedanken* experiments—thought experiments— that he hoped would be able to demonstrate some failure of quantum theory. From these he tried to argue that, if truly indeterministic, quantum theory would lead to catastrophic paradoxes. He further said that to avoid these problems, there had to be some new properties held by each particle-wave beyond those we know about (position, velocity, etc.). Einstein thought that if we knew the values of these other properties, called hidden variables, then quantum physics would act as a fully predictable and deterministic theory. Einstein's objections to the indeterministic quantum theory were faulty, however. Niels Bohr, in particular, let loose blistering attacks at the very foundations of Einstein's arguments. Although remarkably strange and often counterintuitive, quantum theory was self-consistent and free of true paradoxes.

A great deal of confusion that was found among physicists during the early development of quantum theory came from the fact that the quantum world is very different from the classical one with which we are familiar. Although we are able to picture planets orbiting the sun or billiard balls bouncing off one another, we have little or no intuition that is applicable to quantum physics. That lack of familiarity forced (or enabled, depending on how you look at it) quantum physicists to imagine the quantum world for themselves. So although any valid interpretation of quantum physics must agree with all experimental tests, different physicists imagined the quantum world in quite different ways.

In Bohr's interpretation of quantum theory, sometimes called the Copenhagen interpretation, the observer plays a central role. Perhaps the most famous illustration of Bohr's interpretation of quantum physics comes from Erwin Schrödinger, a pioneer of quantum theory, who published an essay describing the famous cat paradox. In this thought experiment, a cat is enclosed in a chamber completely insulated from the outside world, along with a device: a Geiger counter with a small amount of radioactive material that is attached to a cat-killing machine.[4] In the course of an hour, there is some probability (say 50 percent) that a quantum process in the radioactive material will trigger the Geiger counter, which in turn will trigger the cat-killing machine, thus killing the cat. The important point of Schrödinger's thought experiment is that, according to Bohr and the Copenhagen interpretation, up until the point at which someone opens the chamber and observes whether the cat is dead or alive, the cat is neither dead nor alive. Instead, it is in a combination of dead and alive states. Only upon opening the chamber does the mixed quantum

---

4. Although in Schrödinger's original essay, the cat-killing machine is a hammer that smashes a bottle of hydrocyanic acid over the cat, the precise implement of death used by the machine is of no consequence and sometimes varies in different accounts.

state—called the wave function—of the system give way to a single one, and the cat takes on a single value of aliveness or deadness. This transition is often referred to as the collapse of the wave function.

Prior to an observation, then, a quantum object exists in a state that is a combination of different possible physical values: different positions or velocities, for example. Only upon measurement, says Bohr, does the object take on particular values for those quantities.

Bohr's Copenhagen interpretation was by no means the only way physicists thought the quantum world might behave. Despite the fact that Einstein's objections to indeterministic quantum theory had been shown to be invalid, many physicists continued to ask whether it might be possible to restore determinism into quantum theory by including hidden variables. Quantum theories that included hidden variables came in and out of fashion for decades. In the 1950s, in particular, such ideas flourished. Much of the interest waned after 1964, however, when John Bell proved that a quantum theory including hidden variables would not describe the world as accurately as the standard, non-deterministic description of quantum physics. (Bell's conclusions do not strictly apply to all possible theories, so a slight possibility remains that a deterministic quantum theory with hidden variables will supplant quantum physics; so far, no one has been able to formulate one that agrees with the experimental data.)

Introducing hidden variables is not the only way to make quantum physics into a deterministic theory. In 1957, two years after Einstein's death, an American named Hugh Everett III proposed a radical new interpretation of quantum theory that was fully deterministic. Under Everett's interpretation of quantum theory, the wave function never collapses; that is, a particle-wave never takes on one specific value for a given property. Everett did not mean that upon opening Schrödinger's box, an observer would see a cat that was both living and dead. Instead, he said that the observer himself is in a combination of two, non-interacting states, one in which, upon opening the box, he sees a dead cat and another in which he sees a live cat. According to Everett, all quantum possibilities are necessarily carried out and exist in different "worlds." For this reason, Everett's description has become known as the many-worlds interpretation of quantum physics.

An important distinction between the many-worlds and Copenhagen interpretations of quantum physics is the different roles that observers play. In the Copenhagen interpretation, a wave function collapses only when it is observed. The special role that this interpretation holds for the observer is deeply unsettling to me (and to many of my scientific colleagues). It is difficult to imagine why the act of observation would play such a special role within the laws of physics. Furthermore, we cannot actually observe the process of a wave function collapsing. I find the introduction of such an ad hoc and untestable hypothesis troubling.

I imagine that Everett felt similarly about the Copenhagen interpretation. His interpretation avoids the philosophical objection that Bohr's merits by doing away with the collapse of the wave function altogether. Rather than one particle with a combination of positions or velocities, or a cat that is both dead and alive, there are many co-existent worlds both before and after the observation takes place. Before the cat's chamber is opened, there are already many quantum worlds, some that contain a dead cat and others that contain a living one. The act of observation merely informs us as to which type of world we are in. As every quantum possibility is realized in some particular world, the many-worlds interpretation of quantum theory is completely deterministic. This determinism does not exist practically to us; we still cannot predict whether we will see a dead or a living cat when we open the chamber. If we could take the sum of all quantum worlds together, however, all quantum possibilities are always carried out and no indeterminism, or "dice rolling," occurs.

As unlikely as it might sound, one of the best illustrations of the many-worlds interpretation of quantum physics I have ever heard was in a radio commercial for the Minnesota Twins baseball team. In the commercial, which I somehow managed to remember from my childhood, a player for the Twins (the catcher, I think) is talking to a fan. The Twins were not doing well that season, and the player is trying to feel better about this by resorting to quantum philosophizing. He reasons that, according to quantum physics, in some world (or worlds), the Twins are currently leading their division and he is batting .400 with eighty home runs. The commercial's punchline comes when the fan suggests that this means that in some world, he himself is the catcher for the Twins. The player responds (in a break with Everett's physics), "No. There is no world like that."

Although I don't recall thinking about it, I'm pretty sure I didn't take the commercial's view of the Universe very seriously at the time. Similarly, there was a great deal of skepticism toward Everett's proposal when it was presented. Quantum theory had been established for decades, and relatively few physicists were eager to consider radical departures from the orthodoxy that they had been taught. Furthermore, Everett was only a student who had not even earned his doctorate when he introduced his interpretation. Someone as young and inexperienced as Everett was not likely to be taken seriously by many in the scientific community.

In the half century since the publication of Everett's interpretation, however, many scientists have warmed up to his idea. I recently stumbled upon a survey regarding the many-worlds interpretation conducted by L. David Raub, a political scientist. He asked seventy-two "leading cosmologists and other quantum field theorists" whether they thought the many-worlds interpretation was true. Fifty-eight percent agreed, 18 percent disagreed, and the remaining 24 percent were undecided or uncertain.

Regardless of the results of this survey, it is my personal experience that the vast majority of physicists don't think it is their place to judge the different interpretations of quantum theory. The truth is that every experimental prediction that you could ever make using the Copenhagen interpretation is exactly identical to the prediction you would make using the many-worlds interpretation. Because physics is fundamentally an experimental science, with competing theories judged on the merit of their agreement with experimental tests, there is no scientific way of deciding which interpretation is correct. I would go as far as to say that it is not even scientific to suggest that one interpretation is more correct than any other. All the same, it would be hard to find a subject that more physicists love to talk about, especially over a few pints of beer.

• • •

So the sub-microscopic world—full of quantum particle-waves—is awfully weird. Now it is time to consider what all that has to do with the macroscopic weirdness—the dark matter—that is of such importance to the structure of galaxies and the cosmos as a whole. At the time that quantum theory was being developed during the first decades of the twentieth century, there were only a few known types of particle-waves (which from now on I will refer to simply as particles): protons, neutrons, electrons, and photons. Of course, there were also numerous known atoms and molecules, but these were simply combinations of protons, neutrons, and electrons, and thus were not fundamental entities themselves. The remainder of the twentieth century brought discovery after discovery of new elementary particles, leading the discipline of quantum physics to evolve gradually into the field of particle physics.

That string of discoveries began accidentally in 1937, when a particle resembling an electron but about two hundred times heavier was identified. The existence of these heavy electrons, called muons, was completely unanticipated and prompted the physicist I. I. Rabi to exclaim, "Who ordered that?"—a question still in need of an answer today. This was followed by the discovery of the pion in 1947, the kaon in 1948, the lambda in 1951, the delta in 1952, and so on and so on. By 1958 such discoveries were so common that Robert Oppenheimer (the former head of the Manhattan Project, which designed the first atomic bomb) facetiously suggested that the Nobel Prize that year should be awarded to the experimental physicist who *did not* discover a new particle.

Although a long string of discoveries might sound like a windfall for a branch of science, many physicists regarded the new particles as an embarrassment of riches. A theory that included so many seemingly unrelated and random particles did not seem as elegant and powerful as a description of nature with only a few simple ingredients. In frustration, Enrico Fermi once said, "If I had known there would be so many particles, I would have become a botanist rather than a physicist."

Eventually, however, physicists realized that most of the new entities were not fundamental after all, but instead were combinations of only a few truly elementary particles called quarks. A few types of quarks could combine in a large number of ways, accounting for many of the newly discovered particles. Groups of two or three quarks can combine to form protons, neutrons, pions, kaons, lambdas, deltas, and many other composite particles, much like protons, neutrons, and electrons can combine in various ways to generate the entire periodic table of the elements.

With the discovery that a only few quarks made up so many objects, the number of elementary particles known to exist once again became manageable. Along with this, order and patterns began to emerge from these objects' properties. By the 1970s a more complete model, known as the Standard Model of particle physics, had been formulated. The Standard Model describes all the known constituents of nature, has been tested to ever-increasing precision over the past three decades, and has been verified again and again. Every known particle is accounted for, and every interaction predicted by the model has been shown to be correct. The successes of the Standard Model are truly remarkable.

The Standard Model includes two classes of particles: fermions and bosons. Fermions are the particles that we normally think of as matter, or at least its constituents. There are twelve types of fermions in the Standard Model: the electron and the electron's heavier relatives, the muon and tau; the six types of quarks; and the three types of particles collectively known as neutrinos (which I will discuss in the next chapter).

Bosons, on the other hand, are responsible for communicating the forces of nature between fermions. For example, the photon is the boson that communicates the electromagnetic force. Particles with electric charge feel the effects of the electromagnetic force by exchanging photons with each other. Without photons, electromagnetism would cease to act, just as if all of the electric charge in the Universe had disappeared.

Just as the photon results in the electromagnetic force, for each of the other bosons there is a corresponding force. A boson called the gluon, which holds quarks together within protons, and protons and neutrons together within atomic nuclei, communi-

Fermions			
Quarks	Up	Charm	Top
	Down	Strange	Bottom
Leptons	Electron neutrino	Muon neutrino	Tau neutrino
	Electron	Muon	Tau

Bosons
Photon (electromagnetic force)
Gluon (strong force)
W,Z Bosons (weak force)

FIGURE 3.3. The members of the Standard Model of particle physics. The fermions (quarks and leptons) are often thought of as the constituents of matter, while the bosons are responsible for the forces observed in nature.

cates the strong force. There is also a weak force in the Standard Model, which is brought about by two particles called W and Z bosons (figure 3.3).[5]

Such a list of particles might tempt us to ask whether some of them might be found throughout our Universe—and whether some could be the dark matter we are seeking. Many of the constituents of the Standard Model can quickly be discarded as pos-

---

5. You might have noticed that I didn't mention the force of gravity here. This is because we do not currently know how to incorporate gravity into a particle-physics theory. Although we expect that gravity is the result of a boson, called a graviton, that communicates the force, our current knowledge of the particle nature of gravity is limited. Therefore, the Standard Model of particle physics does not include the force of gravity. Also, there is one other boson of the Standard Model that I have intentionally left out of this discussion. I will return to this additional particle in chapter 5.

sible candidates for dark matter, however, because most of them exist only for a tiny fraction of a second after their creation before disintegrating into other, lighter particles. Muons, for example, exist for only a millionth of a second before decaying into electrons and neutrinos. Many of the other Standard Model particles decay even faster. In fact, if at some instant a single example of every known fermion and boson were present, after only one second only the proton, neutron, electron, photon, and neutrinos would remain.

That eliminates a lot of candidates for dark matter. We can eliminate more quite easily. Protons, neutrons, and electrons quickly aggregate to form atoms and molecules. If not confined within an atom's nucleus, even neutrons decay into protons, electrons, and neutrinos in a matter of minutes. But even though protons and electrons are stable, they cannot possibly make up the Universe's dark matter. These objects might cluster together to form various types of MACHOs, but as I demonstrated in the last chapter, MACHOs cannot constitute most of the Universe's missing mass. Furthermore, isolated protons and electrons have a fatal flaw for any dark matter candidate: they have electric charge. Anything with electric charge invariably radiates light. If such particles existed as dark matter, we would easily have detected them. Charged matter cannot be dark matter.

Photons must also be crossed off the list of candidates for the dark matter. In addition to being massless, they are pieces of light and would easily have been detected, and certainly cannot constitute something we would call "dark" matter.

For these and other reasons, we are left with only one class of particles in the Standard Model that could be a candidate for dark matter. If the Standard Model is to provide us with the missing mass of the Universe, the mass must come in the form of neutrinos.

CHAPTER 4

A DARK ANIMAL IN THE QUANTUM ZOO

A particle that reacted with nothing could never be detected. It would be fiction. The neutrino is just barely a fact.

—Leon Lederman

Neutrinos they are very small.
They have no charge and no mass
And do not interact at all.
The Earth is just a silly ball
To them, through which they simply pass,
Like dustmaids through a drafty hall
Or photons through a sheet of glass . . .

—John Updike, "Cosmic Gall," in
Telephone Poles and Other Poems, 1960

I magine you are walking down a city street. You will find yourself surrounded by all kinds of things: other people, buildings, cars, and so on. If you were to walk in a straight line, it would not take long at all before you collided with someone or something. The longer you walked, the more things you would bump into. Your journey is analogous to how a particle of ordinary matter—an electron or a proton, for example—acts within a solid object such as Earth, a table, your body, or whatever. The particle would bounce around, colliding over and over again with whatever surrounded it. The particle might even stick to other objects to form atoms or molecules. Ordinary matter is an active participant, interacting fully with the world around it.

Now picture the city street again. This time, however, imagine that all the other people, cars, and buildings are much smaller, and with vast amounts of space between them. If these potential obstacles were small enough, you could walk in a straight line for a very long time before you would come into contact with anything that could have any influence on your movement. Figuratively speaking, this is the world as seen by a neutrino. A neutrino can usually pass through the entire Earth without interacting with the matter even a single time. It's as if Earth isn't even there. From a neutrino's perspective, everything in the Universe is scarcely there at all. And from the point of view of everything else, neutrinos are almost entirely unnoticeable as well.

Of all the types of matter that make up the Standard Model of particle physics, neutrinos are among the most difficult to study. They are also among the most interesting. These elusive creatures are, in a practical way, disconnected from the world we experience. Billions of neutrinos, traveling at speeds only slightly slower than that of light, pass through your body every second of every day. Nevertheless, because they rarely interact with ordinary matter, you never know it. We can neither see them nor feel them.

Although that elusiveness makes them difficult to study, it also makes them excellent candidates for dark matter.

• • •

Given the elusiveness of the neutrino, it might seem strange that anyone would have thought of such a thing in the first place. The existence of what eventually came to be known as the neutrino was first suggested in 1930, in a desperate attempt to explain the behavior of certain radioactive atoms. It had been known since the late nineteenth century that some elements are radioactive (meaning that they emit energetic particles, known as radiation). Research by physicists such as Henri Becquerel, Marie Curie, and her husband, Pierre, revealed that radiation occurred as three distinct types that were named, in order of discovery, alpha, beta, and gamma rays after the first three letters of the Greek alphabet. Attempts to understand beta radiation would eventually lead to the discovery of the neutrino.

The Curies had shown that radioactive atoms emitting beta rays do so while transforming into slightly less heavy atoms that have more positive electric charge on the nucleus. To balance this increase in positive charge, the beta radiation—which contains energetic electrons—carries away an equal amount of negative electric charge. For example, some types of radioactive potassium atoms become calcium atoms as they emit beta radiation. The original charge is conserved, as the sum of the charges of the emitted electron and the calcium nucleus equals the total charge of the original potassium nucleus. The Curies and other investigators of radioactivity thought that every time the process occurred, the electron emitted (the beta ray) would have the same amount of energy. This is because Einstein's relationship of mass and energy, $E=mc^2$, says that the amount of energy available to be carried away by the electron is proportional to the amount of mass lost in the process, the difference between the masses of a potassium and a calcium nucleus in their example. Because these nuclei each had known masses, the amount of energy freed in such a process could be straightforwardly determined. The emitted electron, by this logic, should have the same amount of energy each and every time this process took place.

However plausible this prediction may have been, the experimental outcome was quite different. The electrons produced as beta radiation did not have identical quantities of energy but instead varied over a range. Worse still, the total amounts of mass and energy apparently were not conserved in this process. Something—or even some principle—seemed to be missing.

When faced with this inconsistency, the scientists of the time were left with only extreme and unattractive solutions. Niels Bohr even went so far as to suggest that in subatomic processes such as beta radiation, the total amount of energy in the system could change. Accepting that hypothesis would have meant throwing out the law of conservation of energy, and would have been as radical a departure from the laws of classical and quantum physics as anyone could imagine.

In 1930 another solution to the beta-radiation problem was suggested. A young Austrian physicist named Wolfgang Pauli, already famous for his work on quantum mechanics, proposed that if another particle were created along with the beta-ray electron, then that other particle could carry away some of the energy. To account for the variations in the speed of the electron and the amount of missing energy, Pauli further proposed that the energy released during each instance could be divvied up differently. Sometimes the electron would carry more of the energy and sometimes less, thus explaining the discrepancies between the theoretical predictions and the experiments.

Pauli first proposed his solution to the beta-radiation problem in a letter he sent to a physics conference on radioactivity in Tubingen, Germany. In his letter, he called these new particles neutrons, although they are not what we have come to know as neutrons. (The particles we call neutrons today, which make up atoms along with protons and electrons, would be discovered by British physicist James Chadwick two years later.) The neutrons of Pauli's letter are what we today call neutrinos.

Pauli's historic, and rather quirky, letter reads in part:

Dear Radioactive Ladies and Gentlemen,

As the bearer of the lines, to whom I graciously ask you to listen, will explain to you in more detail, how because of the "wrong" statistics of the N [nitrogen] and ^{6}Li [lithium-six] nuclei and the continuous beta spectrum, I have hit upon a desperate remedy to save the "exchange theorem" of statistics and the law of conservation of energy. Namely, the possibility that there could exist in the nuclei electrically neutral particles, that I will to call neutrons, which have spin $1/2$ and obey the exclusion principle and which further differ from light quanta in that they do not travel with the velocity of light. The mass of the neutrons should be of the same order of magnitude as the electron mass and in any event not larger than 0.01 proton masses. The continuous beta spectrum would then become understandable by the assumption that in beta decay a neutron is emitted in addition to the electron such that the sum of the energies of the neutron and the electron is constant. . . .

I agree that my remedy could seem incredible because one should have seen those neutrons very earlier if they really exist. But only the one who dare can win and the difficult situation, due to the continuous structure of the beta spectrum, is lighted by a remark of my honoured predecessor, Mr. Deybe, who told me recently in Bruxelles: "Oh, it's well better not to think to this at all, like new taxes." From now on, every solution to the issue must be discussed. Thus, dear radioactive people, look and judge. Unfortunately, I cannot appear in Tubingen personally since I am indispensable here in Zurich because of a ball on the night of 6/7 December. With my best regards to you, and also Mr. Back.

Your humble servant,

W. Pauli

Pauli was clearly uncertain that his proposal was the correct solution to the beta-radiation problem, or even a very good one. He thought his idea deserved consideration only because all of the less-desperate solutions had failed. Indeed, Pauli was so wary of his idea that he did not present his hypothesis in person until three years later in 1933, and didn't publish it in a scientific journal until a year after that.

Others didn't wait so long. By the time Pauli got around to publishing his idea in 1934, Enrico Fermi had developed a complete theory of beta radiation that included Pauli's new particle. In the years between Pauli's letter and Fermi's publication of his theory, Chadwick had discovered and named what we call the neutron, substantially improving physicists' understanding of the atomic nucleus. The discovery of the neutron opened the door to understanding the processes behind radioactivity. Using the new model of the atomic nucleus, Fermi explained beta radiation. He demonstrated that the transformation of a heavier, radioactive atom into a slightly lighter atom happened when an individual neutron inside the radioactive atom decayed into a proton, an electron, and one of Pauli's invisible particles. Because the name neutron was now already taken, Fermi called the new particle the neutrino, Italian for "little neutral one" (figure 4.1).

All interactions between particles are the result of a force. But what force was responsible for the process of neutron decay? When Fermi published his theory of beta radiation, the only force understood within the context of quantum physics was electromagnetism.

Fermi calculated, however, that if the electromagnetic force was responsible for neutron decay, then it should happen billions of times more often than was observed. Therefore, Fermi argued, the beta decay of neutrons had to be caused by a much weaker force. It would be more than twenty years before any direct evidence for neutrinos or Fermi's weak force would be found, but confirmation of both the weak force and the neutrino would come, and Fermi's model would achieve central importance to quantum mechanics and particle physics.

Neutrinos rarely interact with matter. Because of this, extreme measures have to be taken in order to detect the presence of neutrinos. You must have either an enormous, and enormously

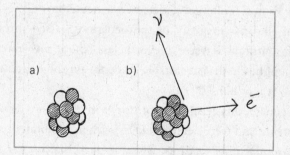

FIGURE 4.1. In some radioactive nuclei, a neutron (white) can spontaneously transform into a proton (shaded) along with an electron (e⁻) and a neutrino (v). The electrons emitted in this process make up what is known as beta radiation. Studies of this process led Wolfgang Pauli to postulate the existence of the neutrino in 1930.

sensitive, detector or an enormous number of neutrinos. Ideally you would have both. By the 1950s very large and very sensitive detectors could be constructed. Now all that was needed were the neutrinos themselves.

The detonation of the first atomic bombs in 1945 horrified the world. But even in terror can be found benign opportunities. Nuclear fission, the process that generates an atomic blast, is so energetic because neutrons decay rampantly. As they do so, they release vast numbers of neutrinos. Nuclear reactors work by utilizing the same fission process, albeit in more controlled circumstances; like the bombs, nuclear reactors generate neutrinos as a by-product of nuclear fission—lots of neutrinos. Perhaps, some physicists hoped, enough neutrinos to be detected directly for the first time.

Although the first experiment to detect the neutrino, conducted in 1953 by Fermi and others, failed to find a conclusive result, in 1959 neutrinos from a nuclear reactor were finally observed. From the nuclear reactor, billions of energetic neutrinos passed though every square centimeter of the experiment's detector every second. Despite this enormous number, the experiment, dubbed Project Poltergeist, detected a neutrino only every few hours as the vast majority of the neutrinos passed through the detector without a trace. Nevertheless, those few interactions were enough: the first irrefutable evidence for the existence of the neutrino was at hand.

Of course, physicists did not stop with the mere detection of the neutrino and were determined to learn more about this strange beast. Among other features, it was learned that there is not just one but three types of neutrino. Also, the particles were found to be nearly massless. And because of their ultra-light nature, they travel at incredible speeds, just slightly under the speed of light.

These ghost-like neutrinos are all around us in vast numbers. Might they be the dark matter of our Universe? They certainly are capable of evading the detection of our telescopes and other astronomical tools. They are also stable and do not decay into lighter objects, unlike most other known particles. The two big questions, then, are: do enough neutrinos exist, and do they have enough mass to account collectively for our Universe's missing matter.

• • •

When the Universe was young and extremely hot, large numbers of neutrinos were created, many of which, we believe, have survived to the present day. Calculations show that there should be, on average, tens of millions of those relic neutrinos in every cubic meter of space in the Universe today. If the relic neutrinos individually had even a very small amount of mass—roughly a hundred-thousandth the mass of an electron—all of the neutrinos in the Universe together could be enough to account for the dark matter.

Until fairly recently, however, no one had been able to determine whether the masses of the three types of neutrino were large enough to make up the dark matter. To many theoretical physicists, the simplest and most elegant scenario was one in which the neutrinos were exactly massless, like the photon. In order for neutrinos to dominate the total mass of the Universe, this argument from elegance would have to turn out to be wrong. But if neutrinos are so difficult to observe and study, how could their minuscule masses be measured?

• • •

The Sun produces enormous numbers of neutrinos as a by-product of the nuclear processes it continuously undergoes. Using a detector the size of a large swimming pool, buried deep underground in South Dakota's Homestake Mine, the first detection of these solar neutrinos was made in 1968. But the most interesting result of the experiment was not that it detected neutrinos from the Sun but rather that it detected too few neutrinos coming from the Sun. Based on the rate at which nuclear fusion takes place in the Sun, physicists had predicted that the Homestake Mine experiment would detect three times more neutrinos than it did. There were three possible solutions to this problem. First, the calculations could be wrong. Second, the experiment could have a flaw, causing it to miss two out of three neutrinos. And third, something might be happening to the neutrinos between the Sun and Earth, causing them to disappear or making them more difficult to detect.

To resolve the problem, the theoretical calculations that predicted the number of neutrinos produced by the Sun were refined, and the experimental equipment in the Homestake Mine was improved and examined to confirm that the measurement technique being used would not underestimate the number of neutrinos coming from the Sun. No significant problems were found in either the theoretical or experimental work that had been done. Something else would have to give.

In the late 1980s and the 1990s, several experiments were carried out to find a solution to the solar neutrino problem. Although the results of those experiments seemed to be consistent with the findings of the Homestake Mine experiment, supporting the conclusion that fewer neutrinos were reaching Earth from the Sun than were expected, many physicists remained somewhat unconvinced. Only in 1998 did the evidence become very compelling that neutrinos produced in the Sun were either not reaching Earth or were somehow changing during their journey. It was in that year that a group of scientists operating an experiment known as Super-Kamiokande announced their results.

The Super-Kamiokande experiment—or Super-K, as it is often called—was essentially a huge tank filled with ultra-pure water buried more than half a mile below Earth's surface. Throughout the 50,000 tons of water in the tank were more than 13,000 detectors capable of detecting incredibly dim flashes of light consisting of as little as a single photon. Ironically, Super-K was not designed primarily to detect neutrinos but rather to look for any sign that protons might be decaying over extremely long times (which is reflected in its full name: the Kamioka Nucleon Decay Experiment). If the protons in the water were decaying, even very rarely, the detectors in the tank were designed to see it. Super-K didn't see any signs of protons decaying, but it did tell us a great deal about neutrinos.

Neutrinos produced in the Sun can easily travel the thousand meters through Earth to the Super-K detector. Of every trillion neutrinos that pass through Super-K, about one collides with a proton or neutron in the water. (Collisions with electrons are even more rare.) Such a collision produces a flash of light similar to the signal expected from the decay of protons. By looking for flashes, Super-K could measure how many neutrinos were traveling through the detector from the Sun.

When the data were studied and counted, it was found that Super-K was seeing only about half of the neutrinos that were expected to come from the Sun. Although this might seem disappointing, it revealed one of the neutrinos' most interesting characteristics—they can transform in flight! Super-K was most sensitive to one of the three types of neutrinos, called electron neutrinos (the other two types being muon neutrinos and tau neutrinos). The reason that Homestake Mine, Super-K, and other experiments were seeing so few neutrinos from the Sun was because some of the neutrinos were transforming during their journey from the Sun into another, less easily detected, neutrino type. This process, called neutrino oscillation, was responsible for the deficit of solar neutrinos that had been observed. Furthermore, a year later in 1998 the Super-K experiment observed that neutrinos generated through particle interactions in Earth's atmosphere were also oscillating. Those two observations were momentous to the particle physics community.

When the Super-K results were announced, I was an undergraduate physics student. Even within the small department at the college I attended, I distinctly remember the excitement that Super-K's discovery produced. None of the faculty were specifically researching neutrinos, but it was clear that they saw the announcement as a historic one. Although I didn't fully understand the implications at the time, it was the first occasion that contemporary physics really seemed exciting to me. I will never forget it.

If neutrinos were indeed oscillating between the three types, then the dearth of electron neutrinos seen by Super-K should be accompanied by a plenitude of those neutrinos as muon neutrinos, tau neutrinos, or both. Confirming this conclusion, therefore, required an experiment that was sensitive to all three types of neutrinos. On June 18, 2001, physicists at the Solar Neutrino Observatory in Canada's Sudbury Mine announced that they had measured the combined number of all three varieties of neutrinos. They found that the total number of neutrinos coming from the Sun matched the predictions made before the era of the Homestake Mine, but that the number of electron neutrinos reaching Earth was once again lower than had been predicted. This result finally confirmed that neutrinos were oscillating from one type into another.

The fact that neutrinos are able to oscillate between different types tells us something very interesting about neutrinos themselves. According to the laws of quantum physics, two types of oscillating particles will transform into each other at a fast rate if their masses are very different, and more slowly if their masses are more similar. If all three types of neutrinos were precisely massless, their masses would be identical and any oscillations between them would be infinitely slow—that is, no oscillations would take place. The observation that neutrinos were in fact oscillating thus revealed that at least two of the types of neutrinos were not massless. The discovery of neutrino oscillations was also the discovery of neutrino mass! And, with mass, the elusive neutrino could indeed be the dark matter of the Universe.

Neutrinos are an example of a dark matter candidate we call a weakly interacting massive particle, or a WIMP. Just like MACHOs, WIMPs are not a single type, but a class of objects. (And just like the name MACHO, the name WIMP is a prime example of the all-too-common physicists' habit of using incredibly childish language to describe their ideas.) Although we have never detected a stable, weakly interacting particle besides the neutrino, it is possible that others do exist. Different types of WIMP might have next to nothing in common. Some might have almost no mass, like the neutrino, or they might be incredibly heavy. They may be slow and gather together in dense clumps, or they may travel at nearly the speed of light in a more-or-less evenly distributed cloud of matter. The possible variety of WIMPs is at least as great as that of MACHOs. WIMPs, however, are much more elusive than the big and clumsy MACHOs. If looking for MACHOs is like looking for a needle in a haystack, then looking for WIMPs is like looking for an invisible needle in a haystack no one has found yet.

• • •

Even if no other types of WIMPs exist, before we name the neutrino as dark matter and march off to take on other questions, we should contemplate what a Universe made up mostly of neutrino dark matter would be like, and compare that with what we actually see when we look at our world. In a way similar to how stars form when clouds of gas collapse under the force of gravity, during the early epochs of our Universe's history, larger structures such as galaxies and clusters of galaxies formed as gravity pulled matter together. Before that period, most of the Universe's matter was spread out more or less evenly across space. For galaxies and clusters of galaxies to begin forming, some locations in space must have been at least slightly more dense than others. The gravity of those clumps would pull surrounding matter toward them, causing the clumps to become denser and denser. Those primordial dense areas would act as the seeds for the growth of structures, eventually producing everything from the largest superclusters of galaxies to individual galaxies, which contain the stars and planets we see today.

How exactly matter gathers together and collapses to form the largest structures in our Universe is difficult to calculate, at least in much detail. To determine exactly how this takes place requires computer simulation. These simulations begin by depicting a large number of objects in space. Given the location and velocity of each individual object at one point in time, the computer program can calculate how and where they will move at later times. The force of gravity felt between the objects perturbs each object's trajectory, pulling them toward the gravitational center of the group. Each object attracts the others to varying degrees, eventually leading to groups of objects becoming bound together into structures (figure 4.2).

Because most of the matter in the Universe is dark matter, the characteristics of dark matter have a great effect on how the Uni-

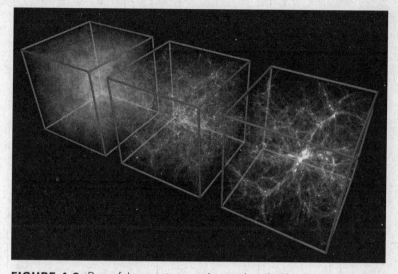

FIGURE 4.2. Powerful computers can be used to simulate how structures such as clusters of galaxies form over time. Here, the distribution of matter in a simulation is shown at three times. The early Universe is shown in the left frame, when very little structure had formed. Gradually, gravity pulls matter together until structures resembling those we see today are present (right frame). These simulations find that if dark matter is hot (moving at speeds near the speed of light) these kinds of structures form too slowly to match observations of our Universe.

Credit: SPL/Photo Researchers, Inc.

verse evolves and on how structures are formed. If the dark matter was made up of slow-moving particles, the simulations found, the dark matter particles would clump together quickly to form dense structures resembling those we see in our Universe today. The presence of slow-moving dark matter, often called cold dark matter, appears to be necessary in order to generate galaxies and other large-scale structures. But neutrinos do not behave like cold dark matter. Neutrinos are not cold. Although they have mass, they are light enough to move at speeds only slightly below the speed of light; this great speed makes neutrinos an example of hot dark matter. Although the computer simulations found that hot dark matter could form large structures, the structures form much too late in the history of the Universe and never become concentrated enough to match the observations of large-scale structures in our Universe. Neutrinos, or any other type of hot dark matter, cannot make up most of our Universe's missing mass.

           • • •

I find it astonishing that by watching for dim flashes of light in subterranean pools of water, seeking to understand something as minute and obscure as the neutrino, we can learn something—actually, quite a bit—about the largest scales of our Universe. In this case we have learned that neutrinos cannot make up most of our Universe's mass. It must be something new and yet to be discovered.

The Standard Model of particle physics contains only one plausible candidate for dark matter: the neutrino. Our observations, along with computer simulations, now tell us that this one candidate is not as plausible as was once thought. Although neutrinos do have the attractive feature of being known to exist, they are no longer thought of as a possible candidate for dark matter. To chase dark matter further we will have to look beyond the Standard Model and toward particles we have so far only imagined.

CHAPTER 5

A GRAND SYMMETRY

The chief forms of beauty are order and symmetry and definiteness, which the mathematical sciences demonstrate in a special degree.

—Aristotle, *The Metaphysics,* Book XIII

The search for the nature of dark matter has led us through the contents of the Standard Model only to leave us empty-handed. Neutrinos once looked like a promising prospect, but no longer. Whatever types of particle or particles constitute the dark matter of our Universe, one thing is known for sure: they are yet to be discovered.

Particle physicists, fortunately, are often good at predicting what types of particles exist before they are discovered. The W and Z bosons, although not discovered until 1983, were predicted to exist by Sheldon Glashow, Steven Weinberg, and Abdus Salam in 1968. The top quark, charm quark, bottom quark, tau neutrino, and others were also predicted to exist before each was ever observed. Theoretical physicists can often predict the likely existence of a new type of particle by studying the patterns of the underlying mathematical structure of the Universe. In some cases, these patterns can lead physicists to notice that a theory appears incomplete, or even leads to a paradox, without the presence of some new component.

As our search has left us with no known candidate for dark matter, we must turn our attention now to the purely theoretical and yet undiscovered. To do this, we will first take a look at the branch of mathematics that guides particle physicists through the deepest and most profound aspects of their theories, potentially enabling them to predict the existence of new particles. We turn now to the science of symmetry.

• • •

If, as Galileo once said, mathematics is the language of science, then symmetry is the dialect of particle physics. Over the last several decades, the mathematics of symmetry has been shown again and again to be among the most powerful tools we have for understanding our Universe and its laws. In many cases, not only do these symmetries help us describe the rules that nature follows, they let us see why these rules are there in the first place.

Everyone has an intuitive understanding of symmetry, but mathematicians and physicists think about it in ways that most of us normally would not. When you think of something that is symmetric, you might think of an image of a vase or perhaps a human face.[1] If you draw a vertical line through the middle of

---

1. Human faces are only approximately symmetric. If you take a picture of an ordinary person's face and replace half of it with a mirror image of the other side, the new picture will seem strange and unnatural.

such a picture, what appears on one side of this line is a mirror image of the other side. Another way of saying this is if you switch or exchange what appears on each side of the line, the overall picture is left unchanged.

That kind of symmetry applies to all sorts of geometric shapes, such as an equilateral triangle (a triangle with three equally long sides). But another kind—a transformational and perhaps slightly less familiar kind—applies, too. In this case, if you rotate the triangle around its center by 120 degrees (one-third of a complete rotation) either clockwise or counterclockwise, the triangle appears exactly as it did before the rotation. This is true not only for 120-degree rotations but for any multiple of this quantity: 120, 240, 360, and so on. The same would be true for a square being rotated by multiples of 90 degrees, or a pentagon by multiples of 72 degrees. Circles contain the most complete version of this kind of symmetry. If you rotate a circle around its center by any angle, it remains unchanged.

Not all examples of symmetry have to do with images or configurations in space, however. Numbers themselves can often display symmetries. For example, if you take the set of all numbers that are integers (. . . −4, −3, −2, −1, 0, 1, 2, 3, 4 . . .) and multiply each number by −1, you are left with exactly the same collection of numbers. Instead of a symmetry in rotation or reflection, the integers possess a symmetry between positive and negative. You will also find that the set of integers is left unchanged if you add or subtract any integer to or from all of them. For example, if you take the collection of all integers and add the number five to each of them, the collection of numbers you are left with is exactly what you started with—the collection of all integers. These examples are just the tip of the iceberg. The integers reveal many other kinds of symmetry, as do many sets of numbers.

Once scientists began to look for symmetries in nature, they found them nearly everywhere they looked. Nature's laws are full of them. Take as an example the law of conservation of energy. According to this law, the amount of energy in the Universe (or in some isolated system within the Universe) never changes. If you know the total amount of energy now, that number will be the total amount of energy at any time in the past or future. This law is a symmetry of nature. Instead of rotating or reflecting something, we are moving it forward or backward in time.

Conservation of energy is by no means a unique example of symmetry within the laws of nature. All conservation laws can be thought of as symmetries. Furthermore, the fact that many constants of nature, such as the speed of light or the charge of an electron, are the same to all observers at all places and times is a symmetry. Over the past several decades, more and more physicists have begun to think about the laws that describe the Universe's behavior in terms of symmetry. The more that physicists have relied on this approach, the more powerful and widely applicable the concept of symmetry appears to be in physics.

Nowhere is the power of symmetry more evident than in the field of particle physics. One symmetry found among all particles is the relationship between matter and antimatter. For each kind of particle that exists, there also exists an antiparticle with the same mass but with opposite electric charge and other quantum properties. The electron, for example, has a positively charged antimatter counterpart called the positron. Neutrinos have antineutrinos, protons have antiprotons, and so on. If all of the matter in the world were suddenly removed and replaced with antimatter, there would be almost no way of telling the difference. Our Universe contains a nearly perfect and very beautiful symmetry between matter and antimatter.

Even more powerful kinds of symmetries, called gauge symmetries, are used by particle physicists. Although physical theories with gauge symmetries were developed in the mid-nineteenth century, the principle of gauge symmetry was largely ignored until the 1950s and 1960s, when it became one of the most important elements—if not the most important element—of modern particle physics.

Gauge symmetry might seem to be an alien concept, but it is actually something just about everyone is familiar with. A common example of gauge symmetry is found every time a bird perches on an electrical wire. Despite the high voltage of the wire, the bird feels no more electrified than if no electrical current were flowing through the wire at all. This is because the electrical current will flow through him only if it can flow from a region of higher voltage to one of lower voltage. If one of his feet were perched on the high-voltage wire while the other was planted on the ground, electrocution would swiftly follow. With both feet on the wire, however, each is in contact with the same voltage, and nothing is felt at all.

To carry this illustration further, imagine that the voltage of everything in the Universe was instantly increased by the same amount. How would we tell that this happened? Well, we wouldn't. Just like the bird on the wire can't tell if the wire has a high or low voltage, we can't identify the absolute voltage of our Universe. In this sense, the absolute voltage scale is just a matter of definition, which we could have equally well set at a larger or smaller value than the one we chose. The gauge symmetry here is that everything behaves the same way regardless of how the scale is set.

The principle of gauge symmetry probably seems cute by now, but I have yet to demonstrate its real power in particle physics: by assuming that a theory contains gauge symmetry, physicists can reveal more about the theory itself. This is true for the theory of electromagnetism, for example. If we write down the equation describing the fermions in the theory of electromagnetism (electrons and positrons, for example), it can be shown that this equation is not, by itself, gauge symmetric. In other words, the theory behaves differently when a different scale is set. To make the theory respect gauge symmetry, another term must be added to the equation. When it is worked out what this new term must be, it is found that it describes a particle exactly like the photon. In other words, the principle of gauge symmetry tells us that if nature contains electrons and positrons, it must also contain photons!

Of course, we already knew that photons exist, so this didn't really tell us anything we didn't already know about our Universe. If we take our mathematical experiment another step forward, however, something new is revealed. In our revised equation of the electromagnetic theory, we have included a photon with exactly zero mass. If we choose, we can write down yet another term in the equation that gives the photon a mass. When we do this, however, we find that once again our theory does not possess gauge symmetry. By imposing gauge symmetry on the theory of electromagnetism, we are thus forced to the realization that photons must have absolutely no mass. If photons had even the smallest mass, this symmetry would be broken, leaving the theory and most of particle physics in serious disrepair. Furthermore, if not for the principle of gauge symmetry, we would have no way of determining whether the photon was completely massless or just very light. By imposing only the most basic and self-evident principles of symmetry on their theories, particle physicists can tell us things we could otherwise never learn about the Universe.

• • •

In the early 1970s physicists in the Soviet Union, including Yuri Golfand, Evgeny Likhtman, and Dmitry Volkov, began working on an entirely new kind of symmetry that related the particles we call fermions to those we call bosons.[2] No direct experimental evidence for this theory then existed, nor does any yet exist today, but its beauty and elegance have made it irresistible to many in the physics community. Under this symmetry, matter particles (fermions) and force-carrying particles (bosons) are intimately connected, unable to exist without each other. The relationship between fermions and bosons, and between matter and force, would eventually become known as supersymmetry.

The supersymmetry hypothesis holds that for each type of fermion, there must be a corresponding boson—its superpartner—and vice versa (figure 5.1). Each particle and its superpartner will have the same amount of electric charge and other properties, as is the case with the symmetry between matter and antimatter. So, just as the electron has its antimatter counterpart in the positron, according to supersymmetry it also has a bosonic counterpart, which we call the super-electron, or selectron. Similarly, the photon has its superpartner, the photino. Neutrinos have sneutrinos. Muons have smuons. Quarks have squarks, and so on and so on.

Standard Model Particle	Superpartner
Electron	Selectron
Muon	Smuon
Tau	Stau
Neutrino	Sneutrino
Quark	Squark

(continued)

---

2. A few years later, these ideas were independently developed in the West.

Top Quark	Top Squark or Stop
Bottom Quark	Bottom Squark or Sbottom
Photon	Photino
Gluon	Gluino
Z Boson	Zino
W Boson	Wino
Higgs Boson	Higgsino

FIGURE 5.1. The supersymmetric partners of the Standard Model particles.

As of the time I am writing this book, not one of these super-partners has yet been observed. That might seem troubling. If supersymmetry were a perfect symmetry of nature, then the electron and its superpartner, the selectron, would have the same mass, and selectrons would therefore be about as easy to produce in particle accelerator experiments as electrons are. That no selectrons have been created in particle accelerators means that selectrons must be far more massive than electrons are. Indeed, this is true for many of the superpartner particles. Because they have not been observed, these superpartners must be much heavier—and therefore more difficult to produce with accelerators—than their Standard Model counterparts. If nature is supersymmetric, it is a broken, or imperfect, supersymmetry; something more like a roughly symmetric human face than a perfect circle.

When I first learned of this situation in graduate school, I wasn't sure what to think of it. Is broken supersymmetry as attractive and compelling a theory as unbroken supersymmetry? Probably not. Does this mean we should abandon supersymmetry as an idea likely to be manifest in nature? Certainly not. Broken symmetries are observed frequently in nature. Even the symmetry between matter and antimatter is slightly broken, for example, with the interactions felt by matter being very slightly different from those experienced by antimatter. Ultimately, this difference is the reason why there is so much matter and so little antimatter in our Universe today. If the symmetry between matter and antimatter were perfect and unbroken, then both would have been produced in the Big Bang in the same quantities and then would have proceeded to destroy each other, leaving no matter behind. It is because slightly more matter than antimatter was produced in the Big Bang that matter exists in any appreciable quantities today. Sometimes a broken symmetry is more beautiful, or at least more practical, than an unbroken one.

Despite the elegance and appeal of supersymmetry, these arguments alone are not enough to determine that it is manifest in nature. Science is science. A valid scientific theory must successfully predict what we see in the world. Perhaps such evidence for supersymmetry will come in the next few years. Perhaps it will take longer. It is also possible that nature is not supersymmetric. Only the experiments of tomorrow will settle the issue once and for all.

• • •

When I introduced the Standard Model of particle physics a couple of chapters back, I mentioned that there was one particle that I was leaving out. I bring it up now because this last particle provides perhaps the most stunning demonstration of supersymmetry's power. The particle is a boson, but it is very different from the photons, gluons, and other force-bearing particles of the Standard Model. Instead of mediating a force in the way we normally think of a force, this other boson, called the Higgs boson, fulfills a different and very important role. It is responsible for the generation of mass.

We normally think of mass as an intrinsic property of matter. Whether we are talking about an electron or a rock, we think of its mass as something we can weigh or otherwise measure—something we cannot change without changing the object itself. Still, the question of why rocks, electrons, and other things have the mass they do remains unanswered. Where does mass come from?

Let's try to forget our preconceptions about the concept of mass for a moment. In modern physics, as we've seen, our intuitions can often be misleading. According to Einstein's theory of relativity, mass is one specific type of energy. Generally speaking, an object has energy either because it is moving or because it has mass, or some combination of the two. In Einstein's theory, there is no distinction between motion and mass in this respect. They are both just types of energy.

To help understand the relationship between motion and mass, imagine two spheres attached to each other by a spring (figure 5.2). If you were asked what the mass of this combination of objects is, you would probably say that it is the sum of the masses of the two spheres and the spring. No more, no less. However, if the two spheres are put into motion toward and away from each other, stretching and compressing the spring, the system would then contain more energy, thanks to the motion of the individual parts. Now imagine that we weigh the system of spheres and spring. According to Einstein's relativity, all energy, whether in the form of motion or mass, behaves in the same way under the force of gravity. Therefore, the energy created by the motion of the spheres moving, along with the intrinsic masses of the spheres and spring, is perceived as the mass of the system. If you were to weigh the system of spheres and spring, you would find that it

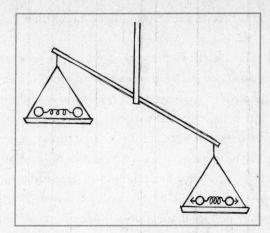

FIGURE 5.2. According to Einstein's general theory of relativity, gravity acts on both mass and other kinds of energy. Therefore, moving objects (such as moving sphares attached by a spring) weigh more than if they were not moving.

weighs more in motion than it does when at rest.[3] Energy in the form of motion feels gravity. Energy in the form of motion generates mass.

Most of the proton's mass is generated in this way—and therefore most of your mass is, too. The proton is made up of three quarks, along with gluons that hold those quarks together. When we add up the masses of these parts, we get a figure that's around 1 percent of the proton's total mass. The vast majority of the proton's mass comes from the motion of the quarks and gluons inside it. Just like spheres held together by springs, quarks held together

---

3. I wouldn't recommend trying this experiment at home. Unless you manage to get the spheres moving back and forth at speeds close to the speed of light, the amount of energy contained in their motion will be very small compared to their mass at rest and you are unlikely to observe any change in the weight of the spheres and spring.

by gluons can generate a total mass much larger than that of their individual parts.

Particles like quarks and electrons, which are themselves fundamental and are not made up of smaller particles, can have their masses generated in another, but not entirely different, way. Quantum theory tells us that particles are constantly popping in and out of existence in the vacuum of empty space all around us. This is one of the consequences of Heisenberg's uncertainty principle, which says that over short periods of time it is impossible to know the precise amount of energy in a system. Particles can therefore be created spontaneously out of nothing as long as they disappear quickly enough. Furthermore, quantum theory not only tells us this might happen, it tells us it must, and is taking place constantly. We are surrounded by a quantum sea of virtual particles.

Included in this particle sea are Higgs bosons. As a particle travels through space, it repeatedly interacts with the ever-present Higgs bosons. Just like with the motion of the spheres and spring, this motion between a particle and the surrounding Higgs bosons adds energy to the particle. The Higgs boson creates the particle's mass. The more a particle interacts with the Higgs boson, the heavier that particle becomes. If a particle does not interact with the Higgs boson at all, like the photon, it can remain massless.

And this process works both ways. Just as Higgs bosons generate masses for many of the particles of the Standard Model, those particles contribute to the mass of the Higgs boson itself. As a Higgs boson travels through space, other types of particles spontaneously emerging from the vacuum constantly contribute to its mass. Considering only the particles of the Standard Model, this process should make the Higgs boson very, very heavy—perhaps as much as 100,000,000,000,000 times more massive than the heaviest particle we have ever observed. The problem, however, is that particle physics experiments have provided indirect, but compelling, indications that the Higgs boson is not in fact extremely heavy, but instead is somewhat lighter than the heaviest known particle. Something, therefore, must stabilize the Higgs boson and prevent it from becoming ultra-heavy. Many theoretical particle physicists believe that the solution to this puzzle lies in supersymmetry.

In the calculation that is performed to determine the Higgs boson's mass, an interesting feature has been observed. The contribution to its mass that comes from a fermion particle popping out of the vacuum is opposite in sign to the contribution from a boson particle: one is always positive and one is always negative. If a given fermion has a boson superpartner, then their contributions to the Higgs boson's mass will cancel each other almost perfectly. This is precisely what supersymmetry guarantees! For every particle that adds to the Higgs boson's mass, its superpartner cancels the effect. The Higgs boson is left with a somewhat large, but completely manageable, mass. Supersymmetry, it seems, is needed to save the Standard Model of particle physics from itself.

• • •

As more and more attention was given to supersymmetric theories, the more successful these theories were found to be. In addition to its ability to stabilize the mass of the Higgs boson, supersymmetry has made the efforts of theoretical physicists to build a more complete and powerful theory that describes all aspects of nature more fruitful.

Throughout the history of physics, many of the most important advances can be thought of as unifications of different aspects of nature that had previously seemed to be unrelated. For example, prior to the work of the nineteenth-century physicists James

Clerk Maxwell and Michael Faraday, the behaviors of electricity and magnetism were thought to be completely separate and distinct phenomena. Electricity was responsible for lightning and electric current, whereas magnetism, completely independently, caused compasses to point north. Maxwell, Faraday, and others observed, however, that a moving or changing electric field was always accompanied by a magnetic field. The two phenomena were eventually found to be two aspects of the same force of nature: electromagnetism.

Since the time of Maxwell and Faraday, Einstein's general theory of relativity has unified the force of gravity with the inertia of bodies. Before Einstein, no one could explain why a heavy object (one that is affected strongly by the force of gravity) is also necessarily an object that requires a lot of energy to accelerate or decelerate. Einstein showed that these two seemingly separate characteristics are in fact inseparable aspects of the same phenomenon. The way that an object's mass and energy warp space-time is responsible for both its gravity and its inertia.

Modern particle physicists dream of a further unification of the laws of nature. The pursuit of a grand unified theory, or GUT, has dominated theoretical particle physics since the 1970s, although the history of the pursuit stretches back much further. Such a theory, it is hoped, would be able to explain why the various particles and forces exist in the combinations they do, and why they have the charges and other characteristics we observe them to have. It is plausible that such a theory could be quite simple, perhaps so much so that it could be written down in a single equation. All of the particles of nature could exist as the consequence of some overarching symmetry. All of the forces within such a theory could merely be different manifestations of the same GUT force.

But what would possibly make us think that the various forces of the Standard Model have anything to do with one another, let alone that they are different manifestations of the same phenomenon? After all, the forces appear to be quite distinct. The electromagnetic force is a whopping 100 billion times more powerful than the weak force. The strong force is a thousand times stronger than the electromagnetic force. Furthermore, the forces affect different particles in different ways. From this perspective, it would appear that these forces have little or nothing in common. But then again, so did electricity and magnetism before they were unified into a single theory.

Appearances aside, the different forces of the Standard Model do have quite a bit in common. The main reason we find ourselves unable to see the similarities is that we study these forces at relatively low temperatures. If the temperature of our environment were a great deal higher—and I'm not talking about Phoenix, Arizona, in July—the strengths of the various forces would also change, leading to a situation in which these forces would become far less different than they might at first appear.

But why is a force more strong or less strong at different temperatures? Consider a single electron. All around it is an electromagnetic field that repels negatively charged particles and attracts positively charged ones. This simple picture is only a convenient approximation of how things really are, however. Remember that in the space of our quantum Universe, pairs of particles are continuously being created and destroyed everywhere. During their brief lives, the positively charged particles in this quantum sea will be pulled slightly toward the electron, forming a sort of cloud around it. The negatively charged particles in the quantum sea similarly get pushed away from the electron. A particle approaching the electron from outside of this cloud does not feel the same electromagnetic field as it would have felt if there had been no cloud around the electron. The cloud of positive particles conceals some of the strength of the electron's electromagnetic field, effectively weakening it. How far the incoming particle can penetrate this cloud is related directly to how much energy it has. The more energy it has, the more of the electron's charge it experiences and the stronger the electromagnetic force appears. Thus the full strength of the electromagnetic force can be felt only by very high energy particles. In the early Universe, for example, when the particles traveling through space were very energetic and hot, relatively little of the electron's charge would have been concealed. At higher and higher temperatures, the strength of the electromagnetic force that is observed becomes larger and larger, approaching its true value.

Similar effects occur for the strong and weak forces of the Standard Model, but with somewhat different results. The strong force, instead of being underestimated, is exaggerated by this effect at low temperatures. Thus as the temperature of the environment is increased, the observed strength of this force becomes weaker. The electromagnetic, strong, and weak forces each evolve differently with temperature.

If we calculate the strengths of the electromagnetic, strong, and weak forces at higher and higher temperatures, we find that they are not very different at all. Only at extremely high temperatures (around 100,000,000,000,000,000,000,000,000,000,000 degrees) do they become equal, however. Even the twenty-seven-kilometer-long Large Hadron Collider (LHC for short)—the most powerful particle accelerator currently planned—is designed only to reach energies a trillion times below the scale of the grand unification of forces. To directly observe the energy of grand unification, an experiment about as large as the entire solar system would be required. For such an experiment to be conducted (or be funded), I'm not holding my breath.

But even if such high energies were attainable, the process of

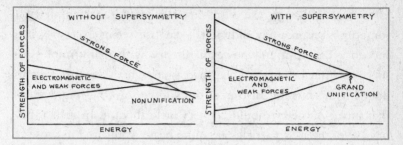

FIGURE 5.3. Without supersymmetry, the three forces of the Standard Model do not evolve to the same strength at a single energy. This poses a major obstacle to developing a grand unified theory. With supersymmetry included, the three forces can converge to a single strength at the same high energy, making grand unification possible. (The electromagnetic and weak forces evolve as mixtures of each other at low energies through a process beyond the scope of this book.)

unifying the forces of the Standard Model would still run into problems. According to calculations that rely on the Standard Model alone, the strengths of the three forces become similar at the unification energy, but they do not become exactly equal. Supersymmetry, however, might again hold the solution. If nature is in fact supersymmetric, the sea of quantum particles all around us contains all the types of particles that exist in nature, including the superpartners. In the early 1990s, calculations done under the assumption that nature is supersymmetric demonstrated that the electromagnetic, strong, and weak forces can have the same strength at a single, very high, temperature. Without supersymmetry, unification is close. With supersymmetry, unification is perfect (figure 5.3).

• • •

I hope you are by now beginning to appreciate why supersymmetry is considered to be so attractive by theoretical particle physicists. Although the ultra-confidence in supersymmetry I sometimes see in a few individual scientists seems excessive, it is indeed a beautiful and powerful theory with many impressive successes. But beauty and elegance aside, what does this have to do with the missing matter of our Universe? One might guess that supersymmetry should say very little about it. In such a theory, all of the new particles are quite heavy compared to those of the Standard Model. In most particle physics theories, heavy particles quickly decay into the lighter particles of the Standard Model, leaving behind any hopes of a viable dark matter candidate. That is, unless something prevents such decay from occurring.

Fairly early in the development of supersymmetry, it became clear that some supersymmetric theories created a series of problems. New interactions introduced by the theories allowed for processes to take place that the Standard Model alone predicted wouldn't happen. Some of these interactions would be seen all around us if they could take place. In particular, this class of supersymmetric theories predicted the decay of the proton.

If it took the average proton just billions or even tens of billions of years to decay (the age of the Universe is about 14 billion years), the results would be catastrophic for our world. Ordinary matter like this book would be disintegrating before our eyes—assuming that there had been any matter left to make a book to begin with. Fortunately, we know that protons do not decay, at least not nearly that fast. Our most precise experiments designed to look for protons decaying find that, on average, protons survive at least 1,000,000,000,000,000,000,000,000,000,000,000 years! That is 100,000,000,000,000,000,000,000 times longer than our Universe is old. The ordinary matter of our world is in no danger of disappearing before our eyes any time soon.

Any model that predicts the rapid decay of protons clearly

must be discarded. This is a problem faced by some models of supersymmetry. Not all supersymmetric models have that problem, however. In some models, an additional symmetry is present that ensures the stability of the proton. Furthermore, this symmetry also makes stable the least heavy of the superpartner particles. So, if we want protons to not decay before our eyes (which we do), it appears that the lightest supersymmetric particle must be stable as well. It is here where we will now look for the dark matter of our Universe.

This new symmetry is called R-parity. Essentially, this symmetry ensures that the evenness or oddness of the number of superpartners is not changed in any interaction. For example, if we collide two ordinary Standard Model particles together in a particle accelerator experiment, the number of superpartners we start with is zero—an even number. In addition to whatever number of Standard Model particles that might come out of the interaction, only an even number of superpartners can be created— whether none, or two, or thirty-two, or some other even number. A single superpartner can never be created in such an interaction. Superpartners can only be created, or destroyed, in pairs.

Much like neutrinos, the superpartner particles are thought to have been created in vast numbers in the early Universe. As the Universe expanded and cooled, these types of particles all decayed into lighter particles, except for the lightest of the superpartner particles. The symmetry of R-parity says that any interaction— such as the decay—of an isolated superpartner particle must leave behind an odd number of superpartner particles. A decaying particle can produce only particles that are lighter than itself, so for the lightest of all the superpartners, there is nothing into which this state can decay. Therefore, it is stable.

So, again much like neutrinos, superpartner particles are all around us in vast numbers—a relic from the Universe's superheated youth. To be a suitable candidate for dark matter, we know, it must be electrically neutral and have no strong interactions. To know whether that is true of the lightest superpartner, we first must determine which superpartner is the lightest.

In the simplest variety of supersymmetric models, there are seven superpartners that are potentially interesting candidates for dark matter. Three of these are sneutrinos, the supersymmetric partners of the Standard Model neutrinos. The other four are the superpartners of the photon, the Z boson, and two Higgs bosons.[4] Those particles, respectively, are the photino, the zino, and two Higgsinos. These four are collectively known as neutralinos. In many supersymmetric models, the lightest of the four neutralinos is the lightest of the superpartners. Which one this might be is not simply answered. Our understanding of neutralinos is complicated by the fact that each neutralino need not be simply one of these four particles, but instead can be some combination of them. For example, one of the four neutralinos could be half the superpartner of the photon and half the superpartner of the Z. Another could be 99 percent Higgsino, half a percent photino and another half a percent zino. Any combination is possible, as long as the four particles add up to a complete set of photino, zino and both Higgsinos.

This is like being told that there are four dogs of four breeds you are looking for: a German shepherd, a beagle, a golden retriever, and a toy poodle. To complicate things, these four dogs can be mixed breeds as long as the sum of the four dogs is precisely one German shepherd, one beagle, one golden retriever, and one toy poodle. There could be four pure breeds. On the other hand, two could be one-quarter German shepherd, one-

---

4. In even the simplest supersymmetric models, five Higgs bosons are needed rather than the one required by the Standard Model.

quarter beagle, and half toy poodle, while another is a purebred golden retriever, and another is half German shepherd and half beagle. The possible combinations are limitless.

With so many combinations possible, the lightest neutralino—our new favorite dark matter candidate—could have a wide range of characteristics. Of the infinite numbers of possible types of neutralinos in all of the possible supersymmetric models, only one set is present in our universe—but which one? The interactions of a Higgsino are not the same as those of a photino, nor are the interactions of a photino the same as those of a zino. Hundreds of studies have attempted to answer questions about neutralinos. How many neutralinos exist in the Universe today? How do these particles interact with themselves or with ordinary matter? And perhaps most important: how can we hope to detect them?

CHAPTER 6

THE HUNT

We've ruled out a lot of suspects and now an arrest is imminent. When you're working on a big case—think JonBenet Ramsey or O.J.—you've got to check out every lead.

—Michael Turner
(regarding the search for dark matter)

It's like Elvis. There are sightings [of dark matter] every so often that are never confirmed.

—Edward "Rocky" Kolb

Deep underground in the Soudan Mine of northern Minnesota is the site of the Cryogenic Dark Matter Search experiment—CDMS for short. The business end of the experiment is a series of silicon and germanium wafers arranged as detectors, each capable of registering the collision of an individual particle of dark matter with itself. Identifying such an impact is not as simple as it might seem. The atoms and molecules making up ordinary matter, such as those found in the germanium and silicon wafers, are constantly bumping into one another, scattering and recoiling over and over again. In such a chaotic environment, an impact by a dark matter particle would be lost in the noise of ordinary motion. It would be like listening for the sound of a pin dropping on the floor at a Metallica concert.[1]

To have any hope of detecting such a tiny impact, we first need to turn down the volume of competing noise. The most important step in achieving this is to make the experiment cold. Very, very cold. The temperature of a material is a measure of how quickly its constituent parts—atoms, molecules, and so on— are moving. (More precisely, the temperature of something is a measure of the average kinetic energy of its constituent parts.) The lower the material's temperature, the less the molecules inside of it are moving, and the less often these molecules collide with each other. At room temperature, the molecules in a piece of ordinary matter bounce off each other at an incredibly high rate. Considering that experiments such as CDMS are hoping to identify dark matter impacts that occur as infrequently as a few times per year, this ordinary motion within the target clearly must be controlled. Cooling the detectors decreases their internal noise. To lower this noise sufficiently, not just cold but incredibly cold temperatures must be maintained.

To talk about such ultra-cold temperatures, scientists generally use an absolute scale, measured in degrees Kelvin. In this

---

1. I'm not trying to suggest that Metallica should be equated with noise. On the contrary, I really like their earlier records.

scale, a room temperature of 65 degrees Fahrenheit is equivalent to about 290 degrees Kelvin. At zero degrees Kelvin, known as absolute zero, the amount of motion within a material is the least allowed by the laws of quantum physics. No lower temperature is possible. The temperature of absolute zero is as close to total silence as can ever be reached. For this reason, the CDMS detector is maintained at a temperature just slightly above absolute zero—about one-hundredth of one degree Kelvin. Its molecules are practically motionless. Reducing the experiment's temperature so much is like listening for a pin dropping on the day after the Metallica concert, when the hall is empty and quiet.

Super-cooling the detector is not enough to enable an experiment to identify individual dark matter impacts, however. Even if Metallica's concert venue were emptied and no sound was being produced inside, noises from outside of the building—sirens, traffic, voices—could still conceal the sound of a dropping pin. To shield your ears from these background noises, the building would have to be soundproofed. Similarly, an experiment hoping to detect dark matter has to be insulated, although the external noise is a little harder to block out than the sound of a siren or a dog barking. The noises that need to be silenced in order to detect dark matter consist largely of the particles from outer space that constantly bombard Earth's atmosphere. Although some of these particles, called cosmic rays, are absorbed before they can reach Earth's surface, many are not. The rate at which cosmic rays would strike a detector is far too high for any signal from dark matter impacts to be identified. To hide from this noisy background, a dark matter detector must be placed far below Earth's surface, where almost no cosmic rays can penetrate. That's why the CDMS experiment is at the bottom of the Soudan Mine, almost half a mile underground. At such a depth, little noise exists to conceal the gentle impacts of dark matter particles.

The CDMS experiment has not yet observed any dark matter impacts. Nor have any of the other similar experiments scattered around the world in places such as France and England. Only one group, running an experiment called DAMA at the Italian Gran Sasso Laboratory, has claimed to have detected the impacts of dark matter particles. Unfortunately, other experiments have refuted that result. Dark matter continues to be undetected.

Despite the silence from dark matter in these experiments thus far, the future is bright. Larger and more sensitive experiments are being rapidly developed. Of all the dark matter experiments worldwide as of 2006, CDMS has reached the highest level of sensitivity. That was accomplished with a detector target weighing only a few pounds. Several experiments are currently being developed that plan to use detectors hundreds of times heavier, and to be more sensitive to dark matter by a similar factor. Furthermore, CDMS may eventually be moved to a deeper mine— perhaps the Sudbury Mine in Canada, which I mentioned earlier in connection with neutrino detection experiments—to lower the level of background noise even further. And detector technology itself is also improving. In the next several years, we may very well see dark matter experiments hundreds or thousands of times more sensitive than those that have been developed so far.

The capacity of these experiments to identify particles of dark matter is not just a matter of technology. It also depends on the nature of the particles themselves. That CDMS and other such experiments have not detected anything so far already rules out some types of supersymmetric neutralinos, for example. Still, of the particles described by all the possible supersymmetric models, the current experiments such as CDMS are sensitive only to a fraction. A much larger number of these models—perhaps the majority of them—predict interactions that are too weak for present experiments to detect but that are within the reach of experiments planned for construction during the next several years. If I were to place a bet about when I think these experiments are most likely to observe neutralino dark matter, I would probably choose a period roughly from 2008 to 2015. Note, however, that I said *if* I were to place such a bet. Unless I were given very attractive odds, I don't think I feel confident enough in this prediction to wager much more than a beer or two.[2]

But what if dark matter is not made up of neutralinos, but of some other particle instead? Does the prediction of this bright future for underground dark matter experiments still hold true? That depends, as you would expect, on the characteristics of the other dark matter particle in question. For some of these particles, the prospects are very good. For others, they are not.

In the previous chapter, I mentioned that in addition to neutralinos there were three other supersymmetric particles that could potentially have the properties we require in a dark matter candidate. These particles, the sneutrinos, are the superpartners of the three neutrinos of the Standard Model. None of the sneutrinos remain popular among particle physicists as a candidate for dark matter, however, because each type would be too easy to detect. If the dark matter of the Universe were made up of such particles, an experiment like CDMS would have seen many thousands of them by now. A rate this high would be impossible to

---

2. Although I don't know of anywhere you can place a bet on the prospects for dark matter detection, for a time you could place a bet with the bookmaking company Ladbrokes on whether the Higgs boson would be found in the next few years. The last time I checked, they were giving 6 to 1 odds against its discovery by 2010.

miss. Experiments have successfully ruled out the possibility of sneutrino dark matter.[3]

Other dark matter candidates could be much, much more difficult to detect. The worst-case scenario is a dark matter particle that interacts only through gravity. In that case, experiments like CDMS (or any plausible expanded or improved version of such an experiment) would never see a dark matter particle. No matter how sensitive and sophisticated dark matter experiments may become, there are no guarantees that they will ever find dark matter.

• • •

Encompassing the Milky Way galaxy is a halo of dark matter. The particles making up this enormous dark matter cloud travel through every corner of our galaxy, oblivious of the planets, stars, dust, and other forms of ordinary matter around them. To a particle of dark matter, the world is a lonely and quiet place.

Very rarely, however, two of these dark matter particles may encounter each other. When this happens, the silence they otherwise maintain can be sharply broken. In the last chapter, I said that neutralinos are stable and do not decay because of the symmetry of R-parity. This symmetry guarantees that the "oddness" of a single, isolated supersymmetric particle is maintained. If two neutralinos come together, however, there is an even number of superpartners present, and R-parity does not forbid them from destroying each other, leaving behind energetic, but otherwise

---

3. I might be exaggerating slightly here. Although in the simplest class of supersymmetric models, sneutrinos have indeed been ruled out as a candidate for dark matter in this way, there are more complicated models in which sneutrinos can still evade these experiments.

ordinary, matter in their place. The ordinary matter that is produced in this way may provide a way of indirectly detecting the presence of dark matter particles.

The process of two neutralinos—or any other type of particle—destroying themselves is called annihilation. In particle physics, however, annihilations don't leave nothing behind. Rather, this process can create many types of matter, all of which ultimately decay into some combination of protons, electrons, photons, and neutrinos. Photons generated in the annihilations of dark matter have a great deal of energy and are called gamma rays. Astronomers have built several telescopes specifically to observe gamma rays, including those from dark matter. Some gamma ray telescopes orbit on satellites and observe gamma rays directly, whereas others are ground-based and observe not the gamma rays themselves but instead the showers of particles created when gamma rays collide with Earth's atmosphere.

Protons and electrons created in dark matter annihilations are more difficult to identify. Although experiments on satellites or on high-altitude balloons are capable of detecting such particles, their signal is overwhelmed by the protons and electrons generated in ordinary astrophysical processes. Unlike the underground dark matter detectors, these experiments cannot be insulated from the background noise. When dark matter particles annihilate, however, they do not only generate matter. They also generate particles of antimatter. For every proton that is created, there is also an antiproton. For every electron, there is a positron. Because ordinary astrophysical processes rarely generate antiparticles, it is much easier for astrophysical experiments to separate the antiprotons and positrons produced in dark matter annihilations from the background noise.

The remaining particle species that is expected to be generated in the annihilations of dark matter particles is the neutrino.

Neutrinos are much more difficult to detect than gamma rays or antimatter particles, however. Even the neutrino experiments I discussed in chapter 4 are not up to the task of detecting neutrinos produced in dark matter annihilations. A location more remote than even the depths of the Soudan Mine is required. To build this neutrino experiment, physicists have had to go to the South Pole.

. . .

Buried about a mile below the polar surface, deep within the Antarctic ice, are the first components of an experiment called IceCube (figure 6.1). Spread throughout a volume of ice roughly half a mile long, half a mile wide, and half a mile high, five thousand highly sensitive detectors are being deployed, each placed in a hole drilled not with a bit but with hot water. When IceCube is complete, it will be the largest neutrino telescope the world has ever seen.

When an energetic neutrino travels near the IceCube experiment, there is a small chance that it will interact with a proton or neutron in the ice. In some of these interactions a muon particle is created, which can then travel farther through the ice and into the volume of IceCube. In a material such as ice, the speed at which light travels is slightly lower than the speed of light in empty space. If energetic enough, the muon will travel at a speed faster than light can travel in the ice, thus breaking the light barrier.[4] And just like a jet creates a sonic boom when it travels faster than the speed of sound, such a muon will create something of a

---

4. This does not violate the special theory of relativity, because the muon never goes faster than light does in empty space, but only faster than light travels in ice.

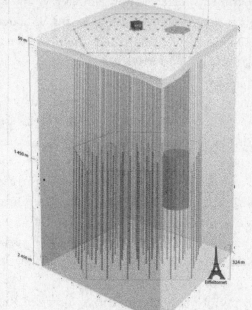

FIGURE 6.1. The IceCube neutrino telescope is currently under construction at the South Pole. When completed, IceCube will consist of thousands of light detectors buried deep beneath the Antarctic ice (up to 2,450 meters deep). Shown for scale is the size of the Eiffel tower.

Credit: The IceCube collaboration.

luminous boom when it travels at such speeds. As the muon zooms through the volume of IceCube, many detectors will register the light of this luminous boom. With that data, the physicists can trace the muon's path and infer the presence of the original neutrino.

IceCube is expected to detect many neutrinos during its operation. Very, very few of these are likely to be from dark matter annihilations taking place throughout our galaxy's halo, however. For studying dark matter annihilations taking place throughout the halo, experiments searching for gamma rays and antimatter are better suited than neutrino telescopes. But although using a neutrino telescope to detect halo-based annihilations is not likely to be fruitful, there is fortunately another strategy for such an experiment to pursue.

As particles of dark matter travel through our solar system, they occasionally bounce off large and dense objects, such as the Sun. When this kind of collision occurs, the dark matter particle typically loses a fraction of its energy, causing it to move more slowly. If slowed down enough, a dark matter particle can become trapped in the gravitational pull of the Sun. Once trapped, such a particle gradually loses more and more of its energy and moves closer and closer to the center of the Sun. There, in the solar core, imprisoned by the Sun's gravity, the dark matter particle remains.

Unlike many of the processes I have discussed, this entrapment is extremely common. In the case of supersymmetric neutralinos, for example, trillions of trillions of these particles are estimated to be captured by the Sun every second. Over the history of the solar system, this adds up to an enormous number of dark matter particles gathered together in the Sun's core. When this much dark matter is gathered together in such a small space, the particles begin to annihilate each other rapidly. Just as in the galactic halo, dark matter annihilation in the Sun's core generates energetic particles: gamma rays, protons, electrons, antiprotons, positrons, and neutrinos. The Sun is so dense, however, that none of those particles can possibly escape—except for the neutrinos. The same characteristic that makes neutrinos so difficult to detect also enables them to pass through the Sun. Upon escaping the Sun, some of these neutrinos—far more energetic than those produced through nuclear fusion—will travel toward Earth. Some of those will travel toward Antarctica. Some of those will even travel

to the IceCube experiment. Of these, a small fraction will interact in the polar ice and be detected.

Neutrinos created by dark matter annihilating in the Sun's
core have not yet been seen. In fact, none of the experiments I
have described in this chapter have yet provided any conclusive
evidence for the particle nature of dark matter. All we have are a
handful of anomalous data that could be interpreted as signals of
dark matter but may turn out to be something far less exciting.
The data could be the product of some other process we do not
perfectly understand, or they could be the result of sheer
coincidence—a few "normal" events appearing in an unlikely
way that resembles the signature expected from dark matter. As
these dark matter experiments are improved and expanded, the
true nature of these hints will become clearer, revealing whether
they are, or are not, evidence for particle dark matter. Although
dark matter has not yet been found, some of these clues are still
very exciting, and may be pointing toward the true identity of
our Universe's missing matter.

• • •

In the summer of 2003, a few months after completing my PhD at
the University of Wisconsin, I was in the process of moving to
my first research position in Oxford, England. Before arriving in
Oxford, I took a detour to the island of Corsica in the Mediter-
ranean Sea, where, in a beautiful seaside village, I attended a two-
week-long series of lectures on particle physics and cosmology.
One of the lectures was given by Joe Silk, the head of the astrophysics department at Oxford University, where I was about to
begin working.

My first conversation with Joe that lasted for more than a few
minutes took place on a boat tour along the Corsican coast. We

began brainstorming a list of ideas that we could work on together at Oxford. Most of our ideas involved searches for dark matter, a topic in which we both are very interested. Among many other things, he mentioned results recently obtained from an experiment called Integral. Integral, a satellite designed to look for gamma rays from outer space, had observed a large number of them from the central region of our galaxy. The origin of the particles was unknown.

Each of these gamma rays seen by Integral contained the same quantity of energy—511,000 electron volts—equivalent to the mass of the electron. This suggested that electrons and positrons were annihilating each other in the inner region of our galaxy and generating the gamma rays. I knew that there were plenty of electrons in the central galaxy for this process to occur, but I had no idea where the positrons were coming from. So I asked Joe precisely this question. His short answer was that he didn't know. Neither did anyone else. The way the data was distributed showed that the positrons were spread out in a sphere a few thousand light years across located at the center of the galaxy. If these positrons were created in supernova explosions, or in other processes involving stars, they should have formed a flatter shape, more like a pancake than a sphere. No conventional explanation could account for the Integral data. But Joe mentioned that another researcher at Oxford, Celine Boehm, had suggested that dark matter particles annihilating throughout the inner region of the galaxy might be able to create the signal reported by Integral.

My first reaction to Boehm's idea was that it must be wrong. The types of dark matter particles that were normally studied, such as neutralinos, would not produce gamma rays with a single quantity of energy like Integral had seen. Instead, particles like neutralinos would generate gamma rays with a wide range of energies, mostly far greater than the energies detected by Integral.

Explanations relying on neutralinos and other conventional candidates for dark matter were completely unable to account for the Integral anomaly.

Neutralinos and other conventional dark matter particles generate gamma rays with too much energy to explain the Integral observation because of their masses, which are too large to release the amounts of energy during annihilation that Integral detected. If dark matter were made up of much lighter particles, however, they could do the trick. To generate the positrons responsible for the Integral signal, these hypothetical particles would have to have masses greater than that of an electron but not much more than roughly a few times greater. Neutralinos are generally expected to have a mass at least 20,000 times or so greater than the electron, and could be as heavy as 2,000,000 times or so the electron's mass. It would take a very different kind of dark matter particle—a much lighter particle with very different types of interactions—to explain this observation.

Celine, it turns out, had been working on exactly this kind of dark matter candidate for quite some time, even before the Integral anomaly was reported. When I met Celine at Oxford, we immediately began working on this idea of hers. It took about a month for us to perform the calculations, and when we were done, I was amazed to see that the idea seemed to work quite well. My initial opposition to this possibility turned out to be the product of studying heavier dark matter particles for so long that I was suspicious of anything outside of this paradigm. I eventually was able to remove my intellectual blinders, but I took quite a bit of convincing before I became comfortable with the possibility of dark matter particles with such small masses. Celine, however, had seemed confident all along that the idea would work out.

About a month after I arrived at Oxford, Celine, Joe, two other collaborators, named Michel Casse and Jacques Paul, and I submitted a short article introducing this idea to the journal *Physical Review Letters*. Almost immediately, a buzz of interest began to form in the physics community. Colleagues would bring up our work at morning coffee, and we started getting invitations to present our results at various conferences and universities. After a few days, a number of journalists even began to call for interviews. Of all the articles I have published, the response to this one was the greatest.

Along with the conventional interest in our work came some more, well, interesting responses. I received one letter from a woman in Chicago written in Lithuanian. I don't read Lithuanian nor do I know anyone who could translate it for me, but looking at the letter I did notice a few words written in English, including "parapsychology" and a reference to an "institute of astrology." I remain perplexed as to why someone in Chicago would write to me in Lithuanian about those topics.

Of all the reactions to our article, the most bizarre—and my favorite—was an open letter we found posted on the Internet. A portion of this letter reads:

> Please tell Dan Hooper and Celine Boehm that we are extremely sorry that they are just now learning about the invisible matter—have they had their heads in the sand for over 5 years?—and needlessly putting them to all this work. For invisible matter was discovered long ago, over five years ago by Slaughter Engineering of Greenwich, CT. Where have you people been keeping yourselves?
>
> Since the rotten US government monitors our every move we make and often destroys it, it will be appreciated if you will please acknowledge receipt of this email IF IN

FACT YOU ARE PERMITTED BY THE US GOV-
ERNMENT TO RECEIVE IT.

If and when you do learn more about Invisible Matter
and find yourselves in our tracks at least 5 years behind us,
please don't be so foolish like some others and claim you
have discovered invisible Matter for you have not. The
very best you can do is to confirm what we discovered
over 5 years ago and are still, every day, learning more
and more about Invisible Matter from the very ones who
created the Universe, our friends and benefactors THE
SUPER INTELLIGENT BEINGS, the SIBs, using what
else, INVISIBLE MATTER.[5]

Signed,

Slaughter Engineering

Immediately after finding this letter on the Internet, I posted
it on the wall of my office. I couldn't make up something as weird
if I tried. I love this stuff.

• • •

Since we published our article on less massive dark matter particles and the Integral anomaly, a number of other scientists have
advanced this idea further, exploring other implications of this
hypothesis and proposing specific models that could generate such
a less massive dark matter particle. We still have not confirmed
whether the signal seen by Integral is the product of dark matter
annihilation or is generated by some other process.

---

5. I have not capitalized these phrases for effect. This is exactly how they appear in the letter.

That is an all-too-common situation in the hunt for dark matter. Clues emerge but no smoking gun is found. Over time, many clues fade away when higher-quality data are collected that don't have the same features, or when some other process is found to be capable of generating the signal in question. Today, a number of very tantalizing hints of particle dark matter are scattered throughout various types of astrophysical data. In addition to the signal seen by the Integral satellite, data from cosmic positron detectors, gamma ray telescopes, and even a CDMS-type underground detector have been interpreted in the last few years as possible indications of dark matter particles. Such clues are exciting but for the moment remain inconclusive. We do not yet know the particle nature of dark matter.

CHAPTER 7

GRAVITY, STRINGS, AND OTHER DIMENSIONS OF SPACE

There may be a whole new universe of large, higher dimensions beyond the ones we can see and every bit as big and rich.

—Joseph Lykken

[The discovery of string theory is like] wandering around the desert and then stumbling on a tiny pebble. But when we examine it carefully, we find that it is the tip of a gigantic pyramid.

—Michio Kaku

Supersymmetric neutralinos make an excellent candidate for dark matter. Numerous search strategies are being deployed to detect these particles, and thanks to them, we will soon confirm that neutralinos make up our Universe's dark matter. Or not.

Supersymmetry is a beautiful and compelling theory, and neutralinos do make excellent candidates for dark matter. But at least for the time being, we have no experimental evidence that these particles exist, let alone that they are the dark matter. If and when that evidence is someday found, I will happily pop the cork on a bottle of champagne and declare neutralinos to be the Universe's missing mass. Until that day, I prefer to keep my mind, and my options, open. Some old saying about eggs and baskets comes to mind.

It should be clear from what I have written so far that quite a number of dark matter candidates have been proposed. To be honest, there are many more than you might imagine. When I am asked to give seminars or conference talks on the topic of dark matter, I like to show a slide I have made with a long list of particles that are thought to be viable candidates for dark matter. Dozens of such possibilities have been proposed over the years, scattered throughout the pages of the scientific journals. Although some of my colleagues—and I, as well—sometimes forget about many of these less-popular dark matter candidates, we should try to remember that the mystery of our Universe's missing mass is not yet solved. Most of the alternative candidates for dark matter are not nearly as appealing as neutralinos, but a few of them are quite compelling. In this chapter, I will describe one of my favorite alternative candidates for dark matter: particles that are the consequence of matter traveling through extra dimensions of space, beyond the three we experience.

• • •

Modern physics is founded on two spectacularly successful theories: Einstein's general theory of relativity and the theory of quantum physics. Neither of these theories has ever been found to disagree with any observation or measurement. They have each been entirely successful in every test that we have applied to them. Nevertheless, at least one of them is not quite right, because the two theories as we currently understand them are ultimately incompatible. The theory of quantum physics describes brilliantly the behavior of our world at the smallest scales—the world of particles, atoms, and molecules. The general theory of relativity describes equally well the behavior of our world in the presence of great quantities of mass and energy—the curvature of space-time generated by stars, galaxies, and the Universe itself. These two theories are each very powerful, but each addresses very different physical situations. Perhaps the greatest challenge in physics today is to understand how these two aspects of nature fit together into a single unified theory—a Theory of Everything.

To call a theory that incorporates both quantum physics and general relativity a Theory of Everything might seem to be an overstatement, but I believe that it is not. Such a theory could potentially include—perhaps even predict—all four of the forces and all of the particles present in nature. Someone with such a theory could use it to derive every piece of information found in all of the physics and chemistry textbooks ever written. The laws describing motion, gravity, electricity, chemical bonding, nuclear fusion, and everything else would be a consequence of this more fundamental theory. The search for a Theory of Everything is the grandest endeavor ever undertaken in the history of science.

Among the best places to search for the key to unifying general relativity and quantum physics are the conditions under which both theories play important roles. If we could compress a large enough amount of energy into a small enough space, perhaps we would witness the effects of quantum physics and general relativity at the same time, and thereby discover how they fit together into a single theory. A particle accelerator would appear to be the likeliest tool to use in such an experiment. Even our most powerful accelerators, however, collide particles with a million billion times less energy than what we expect would be required to see the effects of both quantum physics and general relativity. Present technology does not permit us to study nature under such extreme conditions.

But even if we cannot directly test the laws of physics at such a high level of energy, theoretical physicists still attempt to construct theories that combine the quantum and general relativistic aspects of our world. In such a theory of quantum gravity, we expect the force of gravity to operate in a way similar to the electromagnetic, strong, and weak forces—specifically, gravity would be mediated by a boson, which has been named the graviton. By studying the mathematics of such a theory, we can potentially learn about the properties of the graviton, including how it differs from the other bosons that exist in nature, even if we cannot directly study it in our experiments.

In particle physics, calculations often yield results in terms of probabilities. For example, one might calculate the probability of a particle decaying or of two particles interacting with each other. In processes involving the particles of the Standard Model, these calculations always predict probabilities of such events occurring to be between one and zero—between 100 percent and 0 percent—as we would expect. When physicists began performing analogous calculations with theories of quantum gravity, however, they got some highly unsatisfactory results. Instead of finding "reasonable" probabilities—between one and zero—for the interactions of quantum gravity, physicists found that they were deriving results with infinite, and thus nonsensical, probabilities. In this way, the force of gravity is fundamentally different from, and more perplexing than, the other known forces of nature. Efforts to unify the force of gravity, as it is described by Einstein's relativity, with the theory of quantum physics have been plagued by these kinds of difficulties.

Physicists, of course, have not given up, and continue to search for a unified theory of gravity and quantum physics as aggressively as ever. It is undeniable, however, that the vast majority of those efforts have led to dead ends. It seems that the standard tools used by particle physicists are unable to weld these two very different—and very beautiful—ideas together.

When the available tools are unable to accomplish the task at hand, physicists, like anyone else, begin to search for new tools. These new tools might be quite different from any employed before, and some may take years to learn how to use. Once mastered, the tools may prove not to work how anyone originally imagined. Of all the new tools that physicists have invented in their search for a true theory of quantum gravity, the most promising come from a branch of physics known as string theory.

• • •

String theory first emerged in the 1960s, at a time in which many particle physicists were desperately trying to understand a slew of newly discovered and apparently fundamental particles, including pions, kaons, deltas, and others that I discussed in chapter 3. Today we know that these particles are not fundamental objects themselves, but are each different combinations of quarks. The inventors of string theory, however, knew nothing of quarks, and their theory was an attempt to explain the messy profusion of recently discovered particles. The first incarnation of string theory proposed that many of these particles were not different entities at all, but instead were different kinds of vibrations of the same tiny object, called a string.

String theory suggested that, just as a guitar or violin string can be used to play many notes, a single vibrating string could appear to be many types of particles, their attributes resulting from how the string was vibrating. A string vibrating rapidly, for example, would contain more energy than the same string vibrating slowly, and would thus appear to have more mass. String theorists hoped to account for many of the newly discovered particles as manifestations of only a few strings. They failed in their endeavor, however, and over the next few years a robust quark model for particles was hypothesized and confirmed. Despite this accomplishment, string theory was not totally forgotten, and a few scientists continued to work on the idea. It is not clear what they hoped to find in this seemingly obsolete theory, but their persistence did eventually pay off.

• • •

One of the apparent problems with string theory was its prediction that "extra" particles should exist along with those that had been observed. Many physicists considered the extra particles to be a huge blow to the theory, but others noticed that they had the expected properties of the hypothesized graviton. That was the first indication that string theory might have something to do with how gravity could be combined with a quantum theory in a Theory of Everything. At the time this was first noticed, however, few physicists took this possibility seriously. String theory was still a theory more often ridiculed than revered.

String theory's time came in 1984. In a now famous paper, Michael Green of Queen Mary College in London and John Schwarz of the California Institute of Technology showed that not only could string theory's major problems be resolved, but also that this theory could, in principle, incorporate all four known forces of nature (the electromagnetic, weak, and strong forces, and gravity) as well as all known types of matter. The publication of this paper marked the beginning of what is now called the first superstring revolution, and with it, string theory as a theory of quantum gravity entered the mainstream of science.

As more people became aware of the work of Green and Schwarz, ripples of excitement spread throughout the physics community. Thousands of journal articles on string theory were published over the next few years. To many, it appeared that the dream of a complete and well-understood Theory of Everything would soon be realized. The feeling quickly passed, however, as the equations of string theory proved to be intractably difficult to manipulate. Some progress was made using approximations and other tricks of mathematics, but there were more problems with string theory. After a few frustrating years, string theory was again abandoned by many in the physics community. Just as had happened in the aftermath of the discovery of quarks, however, which appeared to make string theory obsolete, some string theorists stuck with their project, and again their persistence was going to pay dividends down the road. Before that could happen, some big problems had to be solved. One of the biggest was something called the tachyon problem.

A tachyon is defined as a particle that travels faster than the speed of light. That is troubling, of course, because one of the conclusions of Einstein's special theory of relativity is that nothing can travel faster than the speed of light in a vacuum. The Universe has this firm speed limit built into its very fabric. So what would it mean for a tachyon particle to break this rule? Strangely enough, Einstein's theory can be manipulated to show that when something moves faster than the speed of light, in some other frame of reference it is moving backward in time. Tachyons, if they exist, would be very strange creatures indeed.

Any theory predicting tachyonic particles would have serious problems. If particles can travel backward in time, then in principle they could be used to send information backward in time. You can imagine your future great-great-grandchild someday using tachyon-based technology to send a message to the present day instructing someone to assassinate you before you even met your spouse.[1] If the assassination succeeded, the message sender would never be born and thus never send the message and no assassin would kill you, so again your great-great-grandchild would be born and you would be assassinated so they wouldn't be born and then . . . Well, you get the point. Anyone who considers these consequences of particles traveling backward in time—or has seen the movie *Back to the Future*—realizes the logical problems that tachyons present.

The first superstring revolution found a very interesting solution to the problem of tachyons. Namely, it was found that if supersymmetry was included in a string theory, the tachyons present in the theory would naturally disappear. For many theoretical physicists who already loved the beauty and elegance of supersymmetry, this was no price to pay at all. Ever since this discovery, supersymmetry has been considered to be an integral part of string theory—or superstring theory as it had now become.

• • •

The predictions of tachyons and other undesirable particles were not the only reasons that so many physicists had abandoned string theory in its early years. There was another aspect of this theory that appeared so strange and absurd that many discarded it outright. String theory, it seems, is only self-consistent in certain

1. This scenario would make an excellent topic for a futuristic episode of the Jerry Springer show.

cases. For example, in the four dimensions of space-time (three dimensions of space plus one of time) that we know, calculations involving strings suffered from the same failures that other theories of quantum gravity experienced, yielding infinite solutions when only finite ones made any sense. The mathematics of string theory could be used to show that strings could not exist in merely four dimensions of space and time, but only in either ten or twenty-six dimensions. And a realistic string theory—one that actually includes both bosons and fermions—can only be formulated in exactly ten dimensions.

Most people react to this conclusion in about the same way. "What!?" they exclaim. "But we obviously live in a Universe with three dimension of space, right?" Even most physicists responded in a similar way. The idea of living in a world with more than four dimensions of space-time seemed patently absurd, leading many to abandon string theory altogether. Only a few diehards kept a more open mind about this possibility. They hoped to find something that would make these "extra" dimensions different from those we normally experience—something that would make it possible for them to exist without us being aware of their presence. But in what way could they be different? And more importantly, what would this really mean about our Universe? What are the consequences of these extra dimensions?

It turns out that a handy tool for understanding the extra dimensions required by string theory comes from the nineteenth century, when Isaac Newton was the still the undisputed master of physics. This tool is a book written in 1884 by Edwin A. Abbott titled *Flatland: A Romance of Many Dimensions*. The residents of Flatland, including A. Square, the narrator, are members of a species that lives in a two-dimensional world, a mere slice of our three dimensions of space. He recounts for his readers—who "are privileged to live in Space"—how his two-dimensional world is perceived by its inhabitants, and how he became aware of the greater three-dimensional world. Most Flatlanders had no concept of the third dimension of space, and they would have been hard pressed to picture a three-dimensional square (which we call a cube), just as you likely find it extremely difficult to imagine a four-dimensional cube.

Mathematics, however, provides us with some tools to do so. Even if the inhabitants of a two-dimensional world cannot picture a three-dimensional object, it is possible that they could develop a system of mathematics that would enable them to describe it accurately. To a mathematician, the number of dimensions of space is simply the number of variables one needs to describe precisely a specific location. You need only one number to specify a point on a line, and you need two to identify a point on a tabletop. Even if you cannot easily imagine or draw what a four-, five-, ten-, or twenty-six-dimensional object might look like, you can describe objects and spaces with any number of dimensions using the language of mathematics.

Abbott's book is science fiction and not science, but it is remarkably prescient. Nevertheless, *Flatland* does not address one of the most important questions raised by the possibility of living in only some, rather than all, of the dimensions of our Universe. Although it certainly describes what it might be like for beings to perceive only a subset of those dimensions, *Flatland* does not explain why we would not be capable of observing the other dimensions of space. What could prevent us from perceiving all ten or twenty-six of them if they are in fact there? How could they hide from us?

• • •

Like so much in modern physics, the answer to that question takes us back to Albert Einstein. Long before the birth of string theory, physicists and mathematicians had explored the possibility of the existence of extra dimensions of space. Although this subject was generally confined to the fringes of science well outside of the mainstream, it did, at least on one occasion, get the attention of a physicist as prominent as he.

After completing his general theory of relativity in 1915, Einstein spent much of his remaining career searching for a theory that could describe both gravity and electromagnetism in a single unified framework. This turned out to be a challenge too great. At the time, two of the four forces known today—the strong and weak forces—had yet to be discovered. The knowledge of physics that existed during his lifetime was not nearly sufficient for anyone to build a unified theory, even a genius such as Einstein.

Despite ultimately failing to find such a theory, Einstein's search did lead him and other physicists to many interesting ideas. One of the most promising, and most weird, ideas was raised by a German mathematician named Theodor Kaluza. Kaluza explored how Einstein's general theory of relativity would work in five dimensions rather than the ordinary four of space-time. When he worked through the equations, he found that they were mathematically identical to the equations that describe Einstein's theory of gravity in four dimensions and Maxwell's equations of electromagnetism. That bears repeating. Four-dimensional gravity along with electromagnetism is mathematically equivalent to gravity alone in five dimensions.

In 1919 Kaluza informed Einstein of his discovery in a letter asking for the approval of a paper he had written on the subject. Einstein, somewhat astounded by Kaluza's results, didn't know what to think of it. Eventually, Einstein did approve Kaluza's paper, leading to its publication in 1921. Despite Einstein's "thumbs-up," few other physicists took the idea seriously. After all, this might be a neat trick of mathematics, they may have said, but if there were a fifth dimension we would surely see it, right?

In 1926, Kaluza's neat trick of mathematics was shown to be a bit more anchored in the real world than it had been thought. Whereas in the original theory there was no real explanation for how the fifth dimension would remain unobserved, new work by Oskar Klein, a Swedish mathematician, provided an explanation for how it might remain hidden. In Klein's picture, the fifth dimension is "wrapped up" in a circle, something like the way in which the two dimensions of longitude and latitude wrap around a globe. If you walk in a straight line along the equator, you will eventually get back to where you started. Another way to think about how a dimension could be wrapped up comes from a classic arcade game called Asteroids, in which you control a spaceship and try to avoid being destroyed by passing asteroids. For our purposes, the interesting thing about this game is that if you fly off of any side of the screen, you immediately come out on the opposite side. The two spatial dimensions of the game screen are wrapped up in two interrelated cylinders with circumferences equal to the size of the screen. These cylinders are simply projected onto the screen's flat surface; the accompanying figure shows a simplified view of just one cylinder (figure 7.1).

But how, then, does a dimension being wrapped up keep us from perceiving it? Well, in contrast to the three spatial dimensions that we are familiar with, the extra dimension in Klein's theory is very small—about a millionth of a billionth of a billionth of a billionth of a centimeter around! That's small enough to prevent us from becoming aware of the extra dimension's presence.

Although our understanding of quantum physics was still in its infancy at the time of the work of Kaluza and Klein, we know now that the extra dimension described in their model is so small

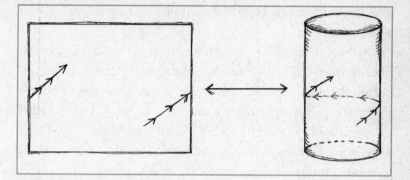

FIGURE 7.1. In the classic video game Asteroids, ships and asteroids that move beyond the edge of the screen simply reappear on the opposite side. This is equivalent to the screen being wrapped up as a cylinder in each dimension. In this way, the geometry of Asteroids is similar to that of the extra dimensions in Kaluza-Klein theory.

that the laws of quantum physics would play a very important role in describing how it would appear to a particle traveling through it. According to the laws of quantum physics, the location of an object cannot be a singular value. So instead of the spaceship in Asteroids existing in precisely one place at one time, quantum effects cause it to be slightly smeared out over space. Because the extra dimension in Kaluza and Klein's scenario is so small, a quantum particle—or quantum spaceship in our analogy—could be smeared out over the entire size of the dimension. It could be everywhere in the extra dimension at once.

The amount of space over which a quantum object gets smeared out depends on its energy. A quantum object with a large amount of energy will have a minuscule wavelength, and thus be smeared out only over a very small volume of space, while a less energetic object would be fuzzier, extending over a larger volume. The smaller an extra dimension of space is, the more energetic an object would have to be to fit into that space; otherwise, its wavelength would be too large. It takes a lot of energy to visit a little dimension.

Most of the particles we measure in our world do not have this much energy. Thus they, like us, experience a Universe that appears to have only three dimensions of space. If we were to build a sufficiently powerful particle accelerator someday, it is possible that the particles accelerated by such a machine could "see" the entire hyper-dimensional span of the Universe. Until then, any extra dimensions of space that may exist will remain hidden.

Even when a particle has more than the minimum amount of energy needed to fit into an extra dimension, it must also have the right wave-like properties to move around a tiny dimension of space. Each time the wave goes around the circle of an extra dimension, the wave patterns of every loop get added together. An infinite number of these waves are added up in this way. If the waves are aligned, or resonate, then the particle-wave can exist in the extra dimension, much like splashing in the bathtub with the correct rhythm can produce satisfyingly big—if messy—waves. But if these waves are not aligned, they interfere with one another, which in the tub leads to calm water and in the extra dimension means no wave, and thus no particle. For the particle to travel in a tiny extra dimension, its wavelength must be equal to the size of the dimension divided by any whole number. The wavelength can equal the size of the dimension, or be $1/2$ that size, or $1/45$ that size—one over any whole number will do. It could not be $1/2.843$—or one over any other non-whole number—of that size without destroying itself.

If there were a particle with enough energy—and the right wavelength—to move through any extra dimensions of space that might exist, what would it look like to us lowly inhabitants of three-dimensional space? Because we cannot see the width of the extra dimension, we cannot tell whether something is moving through it. A particle moving only through an extra dimension would appear stationary to us. Despite this illusion, such a particle is in fact moving a great deal, and would have a large amount of energy in the form of its motion, kinetic energy. According to Einstein's relativity, energy in a form that does not move an object is simply that object's mass. Therefore, to an observer in three-dimensional space, particles traveling through an extra dimension appear to be very massive versions of ordinary particles. These particles traveling through extra dimensions of space, appearing to us to be heavier particles, are called Kaluza-Klein states.

The ideas of Kaluza and Klein ultimately failed to fulfill Einstein's quest for a unified theory of gravity and electromagnetism. For many decades, these ideas were largely abandoned, or at best reduced to a historical footnote thought to have no real physical significance. After this long period of dormancy, the theory of Kaluza and Klein was revived within the context of string theory. Even more recently, physicists have recognized that Kaluza-Klein states might exist all around us. It is even possible that Kaluza-Klein states might make up the dark matter of our Universe.

• • •

In the decade following the first superstring revolution, many more aspects of string theory were calculated and understood. Well, I should say string *theories,* as there was no longer only one theory, but five. Each of these theories, called Type I, Type IIA, Type IIB, Heterotic-O, and Heterotic-E string theory, had its successes as well as its failures. Although each appeared to be self-consistent and was able to avoid the common problems that plagued other theories of quantum gravity, none of these theories appeared to contain all of the phenomena observed in nature. Just as particle physicists felt more confounded by the long series of particles discovered in the 1950s and 1960s, physicists felt more confused than enlightened when the 1980s witnessed the production of so many string theories. After all, there should be only one Theory of Everything.

Of course, the many particle discoveries of the 1950s and 1960s did not end in confusion, but instead led to the discovery of quarks, and eventually to the Standard Model of particle physics. Similarly, the five string theories did not perplex string theorists for too long. In 1995, Edward Witten of Princeton revealed a beautiful solution to this puzzle. The five seemingly different string theories, he showed, were not distinct after all but instead were different manifestations of the same underlying theory. Witten's discovery of this singular theory, called M-theory, ushered in the second superstring revolution. His discovery coincided with the realization that a wide range of hyper-dimensional objects could exist throughout many dimensions of space and time. Much like what happened during the period following the first superstring revolution, the late 1990s witnessed an explosion in string theory research that continues today. Whereas twenty years ago string theory was a highly speculative subject relegated to the fringes of science, today almost every major research university has physicists or mathematicians studying this topic. Although many have tried, it is almost certain folly to attempt to speculate where this path might take us. It will be very exciting to see how string theory evolves and develops over the next years and decades.

• • •

If the picture described by string theory is correct, then in the first instants of our Universe's history, shortly after the Big Bang, the particles present had the energy needed to interact with the world in its full ten-dimensional form. As the Universe cooled and those particles gradually became less energetic, the tiny extra dimensions seemed to disappear. A few of those particles may continue to move through these extra dimensions, however. If those particles are still moving through the extra dimensions today, then they would appear to us—as observers in the seemingly three-dimensional world—to be slowly moving, heavy particles: Kaluza-Klein states. If enough of these states survived, they could be all around us today. Such particles would be a perfect candidate for dark matter. But for them to be dark matter, they must have retained their extra-dimensional momentum throughout the history of the Universe. What would prevent them from gradually cooling and slowing down like the rest of the matter did?

Of all the principles of physics that have been uncovered by humankind, one of the most basic is the law of conservation of momentum. My brother Dylan teaches high school science in a small town in Alaska. In his physics classes, one of the things he tells his students is that momentum is conserved—that something moving cannot slow down or speed up without it transferring momentum (or inertia) to or from something else. A baseball changes direction and moves toward the outfield when it is hit because momentum is transferred from the player's bat. Brakes transfer some of a car's momentum to the road. If you drive your car into a brick wall, your car stops because its momentum enters into the wall.

Roughly the same idea applies to the momentum of a particle traveling through an extra dimension of space. It can lose or gain momentum in the direction of the extra dimension only if it transfers that momentum to or from something else in that dimension. This means that an isolated particle moving through an extra dimension must continue to move through that dimension. A Kaluza-Klein state cannot become an ordinary particle traveling only through three-dimensional space without violating the conservation of momentum. Because of this, Kaluza-Klein states that were created shortly after the Big Bang could have survived throughout the history of our Universe. If enough of these Kaluza-Klein states created in the Big Bang have survived up to the present day, they could be our sought-after dark matter.

If our world's dark matter is in fact the result of particles moving through extra dimensions of space, then the missing mass of our Universe is not really mass at all, but is instead motion hidden thanks to the hyper-dimensional geometry of space and time. The missing mass of our Universe would be essentially an illusion—as would our four-dimensional world itself. The world of our experiences may be very different from the true nature of reality. What we now call dark matter may someday be found to be quite ordinary matter, living in ten dimensions of space-time.

CHAPTER 8

IN THE BEGINNING

In the beginning, God created the heavens and the Earth.

—Genesis 1:1

So long as the universe had a beginning, we could suppose it had a creator. But if the universe is really completely self-contained, having no boundaries or edge, it would have neither beginning nor end: it would simply be. What place, then, for a creator?

—Stephen Hawking

In the beginning the Universe was created. This has made a lot of people very angry and been widely regarded as a bad move.

—Douglas Adams

There is perhaps no topic more widely discussed among theologians, philosophers, and scientists of nearly all times and all cultures than the origin of the Universe. The first sentences of the Old Testament contain the basis of the Judeo-Christian creation story. Although the historical Buddha seems to have had little or nothing to say about it, numerous Buddhist philosophers would later develop detailed descriptions of our Universe's creation. Ancient Vedic writings, the Rig Veda in particular, include multiple accounts of the beginning. Human beings appear to have a universal desire to understand the origin of their world.

Despite the great deal of interest and speculation that the origin of our Universe has generated, science had little to say about it before a hundred or so years ago. The questions of where we, and our world, came from are old hat to philosophers, but are relatively new to the fields of scientific thought. In the nineteenth century biologists began to shed light on the origins of humankind when Charles Darwin formulated his theory of evolution through natural selection. The physicists of the twentieth century have had remarkable success addressing the origin and evolution, not of our species, but of our world. The field of cosmology has become a fully scientific area of inquiry, driven by observational data and rigorous experimentation.

The science of cosmology is integral to our understanding of both dark matter and dark energy. If dark matter consists of particles created in the moments following the Big Bang—which is very likely—then in order to understand this process we must also understand how our Universe has evolved from its form in the first moments following the Big Bang into the world that we see today. If our Universe were unchanging and static, there would likely be no dark matter at all. The matter that inhabits our world, dark or otherwise, is intimately tied to the evolution and history of our Universe.

Our current understanding of dark energy is even more closely connected to our knowledge of cosmology, so much so that it is senseless to discuss the former before the latter. With this in mind, this chapter will be about neither dark matter nor dark energy. First we must visit the origin, evolution, and history of our Universe—that is, the science of cosmology.

• • •

When Albert Einstein set out to develop his general theory of relativity, it was unlikely that he, or anyone else, had any idea that it would reveal the evolution and origin of the Universe. His theory showed that the force of gravity was the consequence of geometry by equating the presence of mass or energy to a degree of curvature in space-time. This relationship would successfully describe not only the orbits of planets and the trajectory of light, but also the geometry and dynamics of the Universe itself.

Because the equations of general relativity are infamously difficult to solve, their applicability has generally been limited to special physical circumstances that exhibit symmetries that enable us to simplify the theory's mathematics. For example, it was possible for Karl Schwarzschild to discover the black hole solution of these equations because of a black hole's perfectly spherical shape: the equally distributed mass results in equal curvature in all directions away from the black hole.

Around the same time, Willem de Sitter, a Dutch astronomer, found the solution for an empty, matter-free universe. Emptiness, as you might expect, is a property that makes the mathematics of general relativity considerably more easy to manipulate. Although our Universe certainly contains matter, the consequences of de Sitter's solution are nonetheless intriguing. De Sitter showed that in an empty universe, space-time would be curved everywhere. Basic laws of geometry, such as the one that states that the shortest distance connecting two points is a straight line, would not be true in such a universe, although they might hold approximately true over short distances. An even more bizarre feature of de Sitter's universe is that it includes a sort of boundary or edge. One can never actually reach this edge, however, because the closer you get to it, the more space becomes curved. This curvature makes distances between points appear to be longer than they actually are. This increasing curvature guarantees that you can never reach this universe's edge, even if you were to travel at the speed of light. In de Sitter's universe, the impression of an infinite volume is created out of a finite one.

Fortunately, even when matter is included, the Universe as a whole also exhibits a special, and very useful, symmetry that makes solving the equations of general relativity possible. Over very large volumes, the Universe is homogeneous—that is, any two locations in space look essentially the same. This might come as a surprise, as we see different things in different places all around us. That is because we experience the world on small scales relative to the size of the Universe. We look at one place in one city, in one country, on one planet, in one solar system, in one galaxy, and so on. If instead we studied a large collection of galaxies, and compared those galaxies to another large collection of galaxies in another part of the Universe, we would find remarkable similarity between the two sets. In the same way, a molecule of fat and a molecule of water are quite different on an atomic scale, but on a human scale, two glasses of homogenized milk look pretty much identical. On the incredibly large scales of cosmology, which deal in collections of clusters of galaxies and even larger structures, all of the fluctuations we see in our experience disappear, leaving behind a very smoothly distributed and homogeneous Universe. This homogeneity allows the often intractable equations of general relativity to be applied comparatively easily to our Universe as a whole.

When the equations of general relativity are applied to a very large volume of space, a peculiar result emerges, namely that the volume of space changes with time. Under the rules of general relativity, space is not a static background, but is dynamic and evolving. Once again challenging our preconceptions about our world, Einstein's theory suggests that the Universe should be contracting or expanding.

If you have not encountered this possibility before, it will probably seem very strange to you—at least it should. Our intuition would lead us to think that in order for something to expand or contract, it must be moving into or out of some larger space. But we are talking about all of space—the Universe itself—expanding. By definition, there is no space outside of the Universe for it to expand into or retreat from. Clearly our intuition does not prepare us for this consequence of relativity. When our intuition fails to handle the strangeness of the physical world, once again I turn to the wisdom of video games.

Recall the arcade game Asteroids (figure 8.1). Imagine the screen has your spaceship in the middle plus some asteroids distributed across the screen. Now imagine that the dimensions of the screen have grown to twice their original size, but that the relative positions of the asteroids have not changed (and, thanks to gravity and electromagnetism holding them together, neither have their sizes). To an observer on the game's spaceship, the asteroids would appear to be moving away from him as the screen became larger—that is, as the Universe expanded. The same would be true of an observer on one of the asteroids. The Universe gets larger without

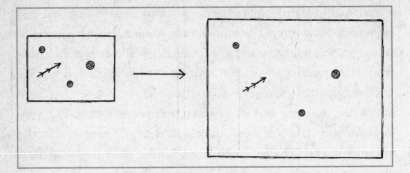

FIGURE 8.1. Returning once again to the video game Asteroids, if the game's screen were to steadily become larger without the ship or asteroids becoming larger as well, everything would appear to be moving away from everything else. Similarly, as a result of our Universe expanding, distant objects appear to be moving away from us.

expanding into anything. And the opposite is true if the Universe is contracting. To an observer, everything in the Universe would seem to be coming toward her, and the distances between all objects steadily become smaller.[1]

In 1922 a Russian mathematician named Alexander Friedmann found the solution to Einstein's equations for a universe containing matter. According to Friedmann's solution, the curvature of a universe's space-time depends on how much matter is present. If there is relatively little matter, gravity's pull upon the universe itself is weak, and Friedmann's universe will expand forever without limit. If there is enough matter, however, gravity will eventually halt and reverse the universe's expansion, causing it to contract. This is not

---

1. If you find yourself asking, "But the game is expanding into something; the expanding screen is expanding into my living room," you can try thinking about it in another way. An expanding screen is equivalent to the asteroids and the ship getting smaller but the screen remaining the same size. As this happens, the distances between two objects appear to get larger when compared to the size of the ship or asteroids.

unlike how throwing a stone from the surface of the Earth results in it falling back, while throwing the same stone at the same velocity from a small moon or space station could cause it to escape and travel away forever.

Einstein was uncomfortable with these conclusions. Perhaps placing too much faith in his intuition and other prejudices, Einstein decided to insist that our Universe was static, neither expanding nor contracting. This was somewhat difficult to accommodate within his theory, however. If there were at least a certain amount of matter in the Universe, then it would expand to a certain maximum size and then begin contracting. If there were less matter in the Universe, it would expand forever without limit. Although neither of those were appealing to Einstein, there was one way for him to wiggle his way away from either conclusion. In the equations of Einstein's theory, there was room to add a component that could counter the force of contraction, potentially balancing it to yield an unchanging Universe. Although Einstein had no empirical reason to add such a component, there also was no mathematical argument that forbade its existence. So, to avoid the unsavory prospect of an expanding or contracting Universe, Einstein included this new component, called the cosmological constant.[2] With this new component fixed to a carefully selected value, Einstein was able to predict a stationary, non-expanding, and non-contracting Universe.

• • •

At the same time that the various solutions to Einstein's equations were being found, observational astronomy was rapidly advanc-

---

2. Einstein originally called this new component the "cosmological term." Only later did it become known as the cosmological constant.

ing as well. New methods allowed astronomers to measure the frequency of light seen by their telescopes. Roughly speaking (as I'm going to ignore some confounding influences such as intergalactic dust on the color of light), if one knows the color of a given type of star or galaxy when at rest, then by looking for the effects of a Doppler shift, an astronomer can deduce whether a distant object is moving toward or away from the Earth. Those getting closer appear more blue while those moving away appear more red. Edwin Hubble and other astronomers began using this technique to measure the velocities at which very distant objects were traveling relative to us. If Einstein was right and the Universe is static, then very distant objects should appear to be, on average, stationary to us just as nearby objects do. That is not what Edwin Hubble found.

In 1929 Hubble announced that all of the most distant objects that he had observed were moving away from us. The Universe was indeed expanding! Although his measurements turned out to be rather inaccurate, his basic conclusions were correct. Today we know that for each light year away from us an object is, the rate at which that object will be receding from us is about an inch or so per second. This is difficult to detect when looking at nearby stars or even galaxies because each object is already moving through the Universe in different directions and at various speeds, and an additional few inches per second just get lost in the wash. If observations are made of objects that are millions of light years away, however, the effect of the Universe expanding becomes far more important, overwhelming the ordinary motion of these very distant galaxies.

Shortly after Hubble's discovery, Georges Lemaître, a Belgian priest and scientist, combined the solution of Friedmann, the observations of Hubble, and his religious convictions to form a new theory. From Friedmann's solution, Lemaître concluded that if our Universe is expanding, then it must always have been expanding. Thinking about this in reverse, Lemaître realized that at increasingly distant times in the past, the more compact and dense the Universe must have been. Furthermore, Lemaître deduced that in the most ancient period of the Universe's existence, there must have been a time at which the density was not only large, but infinite—a singular point in space-time from which the Universe could emerge. This instant would later become known as the Big Bang.

Lemaître, an abbot in the Roman Catholic Church, felt that this new description vindicated his biblical view of creation. The Universe's evolution traced backward to a single point at which, he argued, all creation occurred. He called this point "a primaeval atom"—the creation of God that had grown into our Universe. At first, Einstein strongly criticized the works of Lemaître, believing that he did not properly understand the mathematics of general relativity. Slowly, however, Einstein was won over by Lemaître, Friedmann, and others. Faced with the evidence provided by astronomers such as Hubble, Einstein discarded his static-universe theory, including the cosmological constant factor he had added to it, allegedly calling it the "biggest blunder" of his life.

So as the dust settled, physicists finally had a description of the origin and evolution of the Universe. Those questions that had once been asked only by philosophers and theologians could now be asked—and, more important, be answered—within the context of physics. The science of cosmology had been born.

• • •

If we follow the equations of Friedmann's solution backward in time, as Lemaître did, we can see the rich history of our Universe laid out before us. The further back in time we look, the more hot and energetic our Universe was. When particle accelerators are used to observe collisions at extremely high energies, these machines are in a sense reproducing the conditions of the first moments after the Big Bang. The greater the energy an accelerator can reach, the closer the conditions become to those of the Big Bang. Thanks to such experiments, we have come to understand the series of events and transitions that took place in the very early Universe, and how our world came to be what it is today.

About 14 billion years ago, only the tiniest instant after the Big Bang, the Universe was infinitely compact, infinitely dense, and infinitely hot. Little is known with confidence about this singular moment in space and time. The conditions of our Universe's first instant were so extreme that a full theory of quantum gravity would be needed to understand it. Because we do not yet have such a theory, we know very little about the first tiny fraction of a second, the first tenth of a billionth of a billionth of a billionth of a billionth of a millionth of a second following the Big Bang. So the story I will tell you of our Universe's history begins not at the Big Bang, but at a moment a tiny fraction of a second later.

In its infancy, our Universe was filled with an ultra-hot and ultra-dense plasma of quantum particles. This plasma filled all of space, with all varieties of particles constantly being created and destroyed throughout it. These particles included not only ordinary matter, but also particles of antimatter, supersymmetric particles (if they exist), and likely numerous kinds of other particles that we have not even imagined yet. In these first moments, there were no protons or neutrons. Instead, the quarks and gluons that would someday bind together to form such particles could travel freely. In this earliest form, our Universe contained vast quantities of energy in the form of a dense particle soup.

Even the forces of nature were quite different in these first moments of our Universe's history. There are compelling reasons to believe that in the first fraction of a second following the Big Bang, there were not yet the four forces we now observe in our world. Instead, there was only gravity and the grand unified force. Things did not stay in this condition for long. As these first instants passed, the Universe began to resemble more and more the world we know today (figure 8.2). About a millionth of a second after the Big Bang, quarks began to bind together to form protons and neutrons. As the Universe expanded and cooled, the grand unified force broke into the electromagnetic, strong, and weak forces. Within the first few minutes of our Universe's history, nearly all of the nuclei of the lightest chemical elements, deuterium, helium, lithium, had been forged.[3] By this time, antimatter had long been destroyed. Only a few minutes after the Big Bang, the basic components of our modern Universe were all present. Gradually, these building blocks of our world clustered together to form galaxies, stars, and planets. Nuclear engines inside of stars generated some of the heavier elements of the periodic table, and the explosions of stars generated others. The Universe continued to expand and cool, until it finally reached the point it is at today: a cold and quiet vacuum containing rare bits of matter scattered throughout its vast volume.

• • •

Before a historian presents a description of a series of events, she must first collect evidence in its favor—documents, artifacts, and

---

3. Hydrogen is not listed here is because hydrogen nuclei are simply individual protons and had already formed by this time. Deuterium is a nucleus formed of one proton and one neutron; you may know it as an isotope of hydrogen.

Time after the Big Bang	Event
0	The Big Bang
One millionth of one second	Quarks and gluons combine to form protons and neutrons
A few minutes	Protons and neutrons combine to form deuterium, helium, and lithium nuclei
A few hundred thousand years	Protons and electrons combine to form hydrogen atoms
100 million to 1 billion years	Stars and galaxies begin to form
9 billion years	Our solar system forms
14 billion years	You read Dark Cosmos

FIGURE 8.2. A few of the most notable events in the history of our Universe.

so on—or her argument will be disregarded as mere speculation. Similarly, no matter what Einstein's equations say about it, the theory of the Big Bang and the events that followed would never have become the consensus of the scientific community without the great deal of compelling evidence that has been found over the past century. Experimental evidence and the potential for falsifiability are the cornerstones of all science. Without experimental verification, the Big Bang would be no different from any other creation story or myth.

The first piece of evidence found in favor of the Universe expanding and the occurrence of the Big Bang was Hubble's observation that distant galaxies are moving away from us. In addition to this, the quantities of the light chemical elements—hydrogen, deuterium, helium, and lithium—present in our Universe had been measured. According to the Big Bang hypothesis, these elements were generated in the first few minutes of our Universe's history. In 1948 the calculations of George Gamow and his collaborators predicted for the first time the relative abundances of these elements. Their results agreed with the quantities measured to be present in our Universe, and the abundances of the light elements became the strongest piece of evidence in support of the Big Bang.

Although the observations of Hubble along with the work of Gamow and others appeared to support the conclusion that our Universe had evolved from the very hot and dense state that we call the Big Bang, these fragments of evidence were not yet conclusive. Many scientists, not entirely convinced of the Universe's Big Bang origin, considered alternative theories. In the same year that Gamow wrote his famous paper on the formation of the light elements, Fred Hoyle, Hermann Bondi, and Thomas Gold proposed the most prominent alternative to the Big Bang—the steady-state theory.

The inspiration for the steady-state theory is said to have come from a movie the trio watched together in 1947. After watching the film—a ghost story that ended the same way it had begun—Hoyle and his collaborators recognized that just because something is in motion does not mean that it must ultimately change. Applied to cosmology, perhaps the Universe could be expanding, but without changing in time.

Hoyle, to accommodate Hubble's empirical evidence for an expanding Universe, argued that new matter must be created continuously, repeatedly forming new stars and galaxies at the rate required to keep the overall density fixed. Although matter being created without an equivalent amount of energy being consumed had never been observed—and although the idea contradicted the known laws of physics—it is difficult to prove that such a process could never happen. Furthermore, Hoyle argued that the light elements (deuterium, helium, and lithium) could have been created in stars rather than in the Big Bang. To Hoyle and other proponents of the steady-state theory, the successful prediction of the abundances of the light elements by Gamow and his collaborators was a mere coincidence.

The debate between the Big Bang and steady-state cosmologies raged bitterly for many years. It was Fred Hoyle who first coined the phrase "Big Bang" in an attempt to belittle the competition. Even religion and politics were dragged into the debate. Pope Pius XII strongly declared the discovery of the Big Bang to be affirmation of Catholic theology, much as Lemaître had done. Steady-state theory, in contrast, was often associated with atheism. This rather shallow reasoning was even more ridiculously applied along the political divisions of the Cold War, with Gamow suggesting that the steady-state theory was part of the Communist party line. In reality, the Soviet scientific community claimed that both the Big Bang and the steady-state theory were unsound.

As with all scientific debates, resolution can be reached only through conclusive experimental evidence. Throughout the 1950s and 1960s, pieces of data appeared in favor of each theory, and the question remained an open one. On one hand, the expansion rate had been measured to suggest that if the Big Bang had taken place, then the Universe was only a few billion years old. Because that would make the Universe younger than our solar system, proponents of the Big Bang clearly had a problem.[4] On the other hand, there was little affirmative evidence for the steady-state theory, and comparisons of nearby and distant galaxies favored the Big Bang. The debate wouldn't come to a conclusion until 1965.

---

4. More recent and far more accurate measurements indicate that our Universe is approximately 14 billion years old, which is not in conflict with the age of our solar system.

• • •

In 1960 Bell Laboratories completed the construction of a giant radio antenna in Holmdel, New Jersey. This antenna was part of a system called Echo designed to relay radio signals to distant locations on Earth through a network of high-altitude balloons. In 1962, however, the Telstar satellite was put into orbit and the Echo system became obsolete. With no longer a role to play in communications, the Holmdel antenna could be used for other purposes.

Two radio astronomers working at Bell Labs, Arno Penzias and Robert Wilson, began working with the Holmdel antenna, intending to observe radio waves coming from throughout the Milky Way. They repeatedly experienced problems with their antenna, however; it kept detecting residual noise, or static, which the duo could do nothing to eliminate. At one stage they thought the static might be a man-made background from radio transmitters on Earth, but when they pointed their antenna at what should have been a major source of man-made signals—New York City—they found that the level of background noise did not increase significantly. At another time, they thought that the noise might be the result of the pigeon droppings that had accumulated on the giant antenna, so they cleaned it. The waste was gone, but the static remained. They could not have been very happy.

The more Penzias and Wilson studied this strange background static, the more they became convinced that it was not merely a problem with their instrument. The static came from all directions in unvarying intensity. Eventually, Penzias and Wilson began looking for a less conventional explanation for what their antenna was detecting.

What these astronomers did not know was that two theoretical cosmologists, Ralph Alpher and Robert Hermann, had pre-

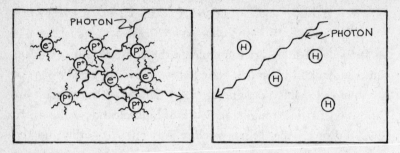

FIGURE 8.3. When the Universe was very young and very hot, electrons and protons moved freely and independently of each other. These charged particles made it impossible for light (photons) to travel undisturbed through space. A few hundred thousand years after the Big Bang, however, the Universe became cool enough for protons and electrons to bind together into hydrogen atoms, allowing photons to travel freely. The photons that were freed at this time are still around us today in the form of the cosmic microwave background.

dicted in 1948 that if the Universe did begin with the Big Bang, then as it expanded and cooled, there would have been a background of light left behind, filling all of space. This background would have been created a few hundred thousand years after the Big Bang, when the Universe first became cool enough, about 4,500 degrees Fahrenheit, for electrons and protons to combine and form hydrogen atoms. Before that transition, photons in this era were constantly deflected and scattered by the charged particles like billiard balls on a full table. Light was essentially trapped in the hot soup of the Big Bang. When protons and electrons joined together, however, there were suddenly very few isolated charged particles left. Only neutral atoms remained. The particles of light were instantly freed, and could travel over incredible distances without deflection (figure 8.3). Alpher and Hermann argued that these photons have been traveling undisturbed since the time of their release, covering a distance of a million billion billion miles, and should still exist today in the form of long-wavelength light called microwaves. They should be everywhere in space and all around us. At any instant, there are about four hundred microwave photons left over from the Big Bang in every cubic centimeter of space, which corresponds to about ten trillion photons passing through every square centimeter every second. Although Penzias and Wilson did not know it yet, it was these photons, now known as the cosmic microwave background, that they were detecting with their antenna.

In their search for an explanation, Penzias and Wilson contacted Robert Dicke, a theoretical physicist and cosmologist at Princeton. Dicke, along with his collaborators, had been unsuccessfully trying to detect the cosmic microwave background for some time. When Dicke was told of the signal Penzias and Wilson had detected with their antenna, he quickly recognized that the two had detected what Alpher and Hermann had predicted more than fifteen years earlier. For the first time, a fossil of the Big Bang—frozen in time for 14 billion years and surrounding everything—had been seen.[5]

The discovery of the cosmic microwave background was the long-awaited confirmation of the Big Bang. The competing steady-state theory had no explanation for the existence of these photons, and was abandoned by all but a few cosmologists.[6] Since Penzias and Wilson's discovery in 1965, the cosmic microwave background has been studied and measured in incredible detail. It is from these measurements that much, if not most, of what we

---

5. You can easily perform your own experiment to detect the cosmic microwave background by tuning your television between channels. Of the static or "snow" that appears on your screen, a few percent of this is the cosmic background from the Big Bang!

6. Although the majority of cosmologists found the steady-state theory to be untenable following the discovery of the cosmic microwave background, Fred Hoyle remained a loyal proponent of the theory up until his death in 2001.

FIGURE 8.4. Arno Penzias and Robert Wilson in front of the telescope they used to discover the cosmic microwave background in 1965. This discovery confirmed that our Universe evolved from an extremely hot and dense state—the Big Bang.

Credit: Courtesy Lucent Technologies.

know about our Universe's history has been learned. It is for good reason that Penzias and Wilson won the Nobel Prize for Physics in 1978 (figure 8.4).

• • •

Cosmologists' interest in the cosmic microwave background certainly did not end with its discovery by Penzias and Wilson in 1965. In the ensuing decades, other astronomers detected the radiation using different ground-based telescopes and techniques. These efforts did verify the earlier findings, but relatively little else could be learned in this way. Earth's atmosphere distorted the signal, and only parts of the sky could be seen by a given ground-based experiment. To overcome these obstacles, the cosmic microwave background would have to be observed from space.

On November 18, 1989, the Cosmic Background Explorer—called COBE, for short—was launched into orbit. Although previous experiments had detected the cosmic microwave background, COBE was the first to measure it precisely. In 1990 COBE scientists announced that they had measured the temperature of this background to be 2.725 degrees Kelvin (about −454.7 degrees Fahrenheit), only a few degrees above absolute zero. The most interesting part of their findings was not the temperature, but rather that it was the same temperature in all directions. The cosmic microwave background appeared to be remarkably uniform, which indicates that the Universe was remarkably uniform at the time when the background radiation originated, a few hundred thousand years after the Big Bang.

Cosmologists knew, however, that the cosmic microwave background, and the Universe it came from, could not be absolutely uniform. If it were, then as the Universe had expanded, all of the matter in it would have spread out evenly, galaxies would never have formed, and our world would look nothing like it does today. If there were even the slightest nonuniformities in the density and temperature of the early Universe, however, then the densest points in space would attract nearby matter and eventually lead to the formation of galaxies and other structures in our Universe. The scientists working on COBE were hoping to see these primordial nonuniformities—the "seeds of structure"—in the cosmic microwave background. In 1992 they announced that they had.

As COBE collected more and more data and measured the cosmic microwave background with increasing accuracy, scientists began to identify tiny fluctuations. Around the average temperature of 2.725 degrees Kelvin, the cosmic background had variations of plus or minus a few thousandths of a degree or so in different parts of the sky. The slightly hotter and colder parts of the sky correspond variations in density that eventually led to the formation of clusters and superclusters of galaxies. The least dense points are those locations that eventually became the vast voids of our Universe—empty space stretching out between the largest structures of matter. The cosmic microwave background had embedded into it a sort of blueprint of cosmic structure, a map of our Universe over the largest scales (figure 8.5).

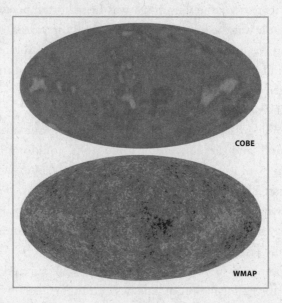

FIGURE 8.5. The temperature of the cosmic microwave background in different parts of the sky as measured by the COBE experiment and the WMAP experiment.

Credit: NASA/WMAP Science Team.

• • •

With the discovery of the cosmic microwave background by Penzias and Wilson, the theory of the Big Bang became the accepted description for the evolution and history of our Universe. With the measurements of the COBE satellite, we entered into what is often called the era of precision cosmology. Whereas the previous measurements of the light element abundances and the discovery of the cosmic microwave background were crucial confirmations of the Big Bang, COBE and the experiments to follow have gone much further. In this era of precision cosmology, cosmologists are not only providing confirmation of the Big Bang theory, but actually uncovering the detailed properties of our Universe. Some of these observations have confirmed what we expected to be true. Others have revealed far less anticipated results, revealing strange new aspects of our world. Among other discoveries, the era of precision cosmology has revealed the existence of what has become known as dark energy.

CHAPTER 9

THE WEIGHT OF EMPTINESS

By convention sweet, by convention bitter, by convention hot, by convention cold, by convention color: but in reality atom and void.

—Democritus

Of all the discoveries of the age of precision cosmology, perhaps the strangest has to do with the nature of empty space. At first thought the idea of empty space seems unremarkably simple; compared to what we know about matter and its motion, surely a description of the nothingness of space must be trivial. Not so. In fact, not only is the nothingness of space not as simple as it seems, but contrary to appearances, space is not nothingness. Even the vacuum of space is never truly empty.

In the first chapter of this book, I mentioned a few of the ideas that early philosophers had concerning the stuff things are made up of. These thinkers were not only interested in matter, however. In the fifth century BC, for example, Democritus and his teacher Leucippus (who argued that matter was made of indivisible atoms) began thinking about the nature of space as well. They described the world as a collection of atoms distributed over an empty background—a void. That seems simple enough, but even their simple description led them to many interesting questions. They wondered whether absolute locations in space exist, or whether the location of one point can be defined only in relation to another point. And they wondered, if one location in space exists only in relation to another, could space exist without atoms inside of it—that is, does space exist in and of itself, or is it merely a concept we have invented in order to describe the motion of objects in our world?

Philosophers have been offering answers to those questions ever since. Aristotle "solved" the problem of the relativity of space by asserting that the void had a boundary. According to his picture of the Universe, a great celestial sphere encompassed all of space, and nothing, not even the void, existed outside of it. The celestial sphere provided an absolute frame of reference that enabled Aristotle and others to think of space as real.

The scientists and philosophers of the Scientific Revolution and the Enlightenment had their own solutions as well. Perhaps the most influential was Isaac Newton's. The success of Newton's theory of gravitation and laws of motion had elevated him to the status of the world authority on such subjects. From this platform, he argued in favor of the existence of an absolute space and absolute time, independent of any measurement, motion, or even the presence of matter. Despite the objections of many of his contemporaries, including Gottfried Leibniz and George Berkeley, Newton's metaphysical position of absolute space and time became the leading doctrine of the age.

It was the nineteenth century before the flaws in Newton's

position were made evident. Ernst Mach, an extremely rigorous-minded physicist with a strong contempt for untestable philosophical positions, correctly recognized that the only measurable distances are the relative ones between objects. There is simply no way to measure an absolute location in space. Imagine that I claimed that the Fermi National Accelerator Laboratory, where I work, could be found at these absolute coordinates in our four-dimensional space-time: 1, 59, 3000, 6. Mach would ask how I knew that, and what measurement could be performed to confirm it. Through measurement, I can determine that the laboratory is about thirty-five miles west of Chicago, or that I was working there 2005 years after the first date on the Gregorian calendar, but I cannot determine any point in space or time without reference to another. Mach would have rejected my claim to an absolute location just as he rejected Newton's concept of absolute space and time. All that can be measured is the relative motion of objects. As a result, Mach argued, all laws of physics must be formulated without any absolute frame of reference—an idea that would have an enormous influence on Einstein. Although it was Einstein who said of absolute space, "It conflicts with one's scientific understanding to conceive of a thing which acts but cannot be acted upon," it was clearly the ideas of Mach that led him to this conclusion.

From that springboard, Einstein leapt to his theory of relativity, in which space—or as he formulated it, space-time—is not only not absolute, but is dynamic, responding to the presence of mass and energy. Einstein's conception of an active and mountainous space-time—not at all the flat and forever unchanging background of Newton—brought into question many of our preconceptions about space. The laws of geometry, for example, had been taken for granted; thanks to Einstein, physicists recognized them to be the consequence of the properties of space-time. Geometry, like space-time, can be different at different locations in space and time.

While Einstein was demonstrating that space was dynamic and not absolute, quantum physics was undermining the notion of the void. One of the implications of Heisenberg's uncertainty principle is that even if no matter is present in the Universe, particles can materialize out of nothing as long as they exist for only a very short length of time. Small fluctuations in the amount of mass or energy in a system are possible. The Universe allows them to be borrowed as long as the loan is promptly paid back, a condition that is satisfied once the particles annihilate one another and return to the apparent void. This process is happening constantly and everywhere. At any given moment, these particles, called virtual particles, are present in every piece of empty space. Which is to say there is no such thing as truly empty space.

If energetic particles can spontaneously emerge from space, then it would seem reasonable that space itself might contain energy or have mass. Assuming that's true, then it is obvious that space, rather than being the inert background that Democritus envisioned, not only would respond to the matter within it, but could have profoundly influenced the history, evolution, and destiny of our Universe.

• • •

One of the predictions that Einstein's general theory of relativity makes is that the Universe's geometry, or shape, can come in one of three basic types. Which form our Universe takes depends on the quantity of energy and matter within it. If a universe contains a sufficiently high density of matter and energy, then it will be curved in on itself, which cosmologists call a closed universe. The surface of a sphere is an example of a closed space in two dimensions. If our Universe is closed, then its geometry would resemble that of a sphere's surface, but in three dimensions instead of only two. The essential point is that any two lines that appear parallel in a closed universe will ultimately converge, much like lines of longitude do at the poles of Earth.

It would seem obvious that if a closed universe is like the surface of a sphere, it must be possible to travel far enough along a straight line so that you would find yourself back at the point where you started. There is a complication, however. If a universe is expanding, as ours is, traveling around it becomes more difficult. Imagine traveling around Earth, and that as you did so, Earth was expanding like a balloon steadily inflating. In such circumstances, the total distance of the trip would get longer as time passed. If this imaginary Earth were expanding rapidly enough, then the distance you would have to travel would grow faster than you could travel, and you would never complete your trip. Fine, you might say, I can walk at only four miles per hour, but if I could travel much faster—say at the speed of light—surely then I could outrun the expansion of Earth, or the Universe. Well, as it turns out, even if you could travel through space at the speed of light, it is possible for a universe to expand faster—the "speed limit" of the theory of relativity doesn't apply to the expansion of a universe—and you would be unable to complete a trip around it.

Space being curved does not guarantee that a universe must be closed. If there is a low density of matter and energy in a universe, then it will be curved such that, instead of closing in on itself, it continues on forever with infinite extent. If you follow two parallel lines far enough in such an open universe, the lines will slowly drift away from each other, no longer remaining parallel. The curvature of an open universe is like the surface of a horse's saddle in three dimensions. As you follow two lines that are parallel at the center of the saddle toward the pommel, the lines gradually spread apart.

Whether our Universe is open or closed depends on how much energy and matter it contains (figure 9.1). If the density is greater than a specific value or less than that value, space-time will be closed or open, respectively. The amount of mass and energy that separates these two possible geometries is called the critical density, and is about one gram per hundred trillion cubic kilometers of space. This may seem like a quantity too small to be of any consequence—water, after all, is more than a billion, trillion, trillion, trillion, trillion times more dense—but keep in mind that the vast majority of our Universe is not filled with water, stars, planets, or even dust, but is mostly the (nearly) empty

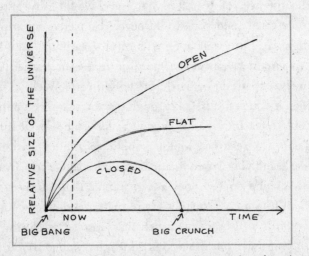

FIGURE 9.1. The history of the expansion and contraction of a universe depends on whether it is open, closed, or flat. An open universe will continue to expand forever, whereas a closed universe will reach a maximum size and then begin contracting, ultimately ending in a big crunch. A flat universe will expand increasingly slowly, gradually coming to a halt.

vacuum of space. The actual density of our Universe that has been measured, in fact, is not too far off from this critical value.

This brings us to the third possible geometry for a universe. If there is precisely the critical density of matter and energy in a universe, then there will be too little matter and energy for a closed geometry, but too much for an open one. In such a universe, parallel lines neither converge nor diverge—they remain parallel, and the laws you were once taught in high school geometry are valid. Such a universe is called flat.

In 1995, when I began studying physics in college, nobody knew whether we lived in an open, closed, or flat universe. Measurements had been made of the average density of matter over very large volumes of space, but they lacked the necessary accuracy to reveal the geometry of our Universe. That was soon to change.

On April 26, 2000, a collaboration of scientists who had been conducting an experiment called BOOMERANG announced its findings. BOOMERANG (which is short for Balloon Observations of Millimetric Extragalactic Radiation and Geophysics) was a detector flown on a high-altitude balloon around the South Pole. During its ten days in flight, the detector made a billion measurements of the cosmic microwave background. The measurements gave physicists a far more nuanced map of temperature fluctuations in the background than COBE had done. The fine-scale nuances in the cosmic microwave background are of great importance to cosmologists. If the sizes of these hot and cold spots could be accurately measured, then the geometry of our Universe could finally be determined.

The results of BOOMERANG showed that among the temperature fluctuations in the cosmic microwave background, the most prevalent were on average about three-quarters of one degree of arc across—the value predicted for a flat universe. A few years later, a satellite-borne experiment called WMAP (the Wilkinson Microwave Anisotropy Probe) measured these fluctuations even more accurately. Once again, the measured density of our Universe was consistent with the value of the critical density needed for a flat universe. Nearly one hundred years after Einstein's first paper on his theory of relativity, the question of our Universe's geometry had finally been answered.

Of course, every good answer in science raises ten more questions. Prior to the age of precision cosmology, conventional wisdom among cosmologists held that if a universe was flat, then it would have just enough gravity so that as it expanded, the rate of expansion would gradually decrease but never quite stop. Rather, it would just coast along at ever-slower speeds. This undoubtedly would have become the standard picture of our Universe's evolution and destiny if not for one thing. Our Universe does not appear to be slowing down at all. It is accelerating.

• • •

As I discussed in the last chapter, objects throughout the Universe are moving in all sorts of directions and at all sorts of speeds. This makes it difficult to know whether something is moving away from you because the space between you and it is expanding, or just because it is randomly moving in that direction. Only for very distant objects is the rate of expansion large enough to dwarf the ordinary random motions of astrophysical bodies. So to measure the rate at which our Universe is expanding, astronomers study very distant objects, just as Hubble did in the 1920s. But the objects measured by Hubble—galaxies—are not very useful to modern astronomers studying the rate of expansion. To study the expansion rate of our Universe over a substantial fraction of its history, another kind of exceptionally bright object is needed.

Exploding stars known as supernovae are to modern cosmologists what galaxies were to Hubble.

One class of supernovae is particularly useful to cosmologists. Whereas most types of supernova explosions occur when a star runs out of nuclear fuel and collapses under the weight of its own gravity, one type, called a type Ia supernova, comes about from the interaction of a pair of stars orbiting each other. If a pair of unequally sized stars are close enough to each other in their orbits, gravity can slowly tear matter from the smaller star and pull it onto its more massive partner. Eventually, the larger star will grow so large that the pressure from its nuclear fusion can no longer support its mass. When this limit is exceeded, the star collapses and then explodes as a supernova.

The maximum mass such a star can support is called the Chandrasekhar limit, named after Subrahmanyan Chandrasekhar, who made the pioneering calculations for such phenomena in the 1930s. The Chandrasekhar limit is about 1.4 times the mass of our Sun. A massive star slowly devouring its smaller companion is the ultimate example of cosmic gluttony, eating more and more until it finally explodes.

Because all type Ia supernovae have approximately the same mass when they explode—that of the Chandrasekhar limit—they also release about the same amount of energy. This is what makes type Ia supernovae so useful to cosmologists. Other types of supernovae explosions can be produced by stars with a wide range of masses, and thus each can release a different quantity of energy. Because each type Ia supernova releases the same quantity of energy, each has the same intrinsic brightness. Astronomers call such objects standard candles. Without them, we would know far less about the evolution of our Universe.

Imagine you wake up one morning to find that Earth is expanding. Of course you want to determine the rate at which

Earth is expanding, so you begin measuring the speed at which two points on the surface are receding from each other. To do this, you point your telescope toward a source of light somewhere else on the planet—let's say the light at the top of a distant light-house.[1] By measuring the Doppler shift of light from the light-house, you can determine the velocity at which it is moving away from your location. Let's say it's a hundred miles per hour. That's great, but it's not enough information to determine how fast the planet is expanding. To determine the expansion rate, you need not only the velocity between two points, but also the distance separating those two points. Somehow, you need to find out how far away from you the lighthouse is. This is where standard candles come in.

Different lights, like most types of supernovae, have different degrees of brightness. But imagine you call the lighthouse keeper and tell him to shine a ten-kilowatt light in your direction. Now because you know the intrinsic brightness of the light source, you can infer how far away it is by measuring its apparent brightness. By combining the speed at which the light recedes with the distance it is away from you, you can determine how fast the rate of expansion is. And what works in this fantasy of an expanding Earth works exactly the same way with type Ia supernovae in space. With them, we can determine the rate at which our Universe has been expanding over much of its history.

In the 1990s cosmologists across the world sought out distant type Ia supernovae. They did this using many of the world's most powerful telescopes, including the Keck Telescope in Hawaii, the Cerro Tololo Inter-American Observatory in Chile, and even the Hubble Space Telescope. Over the course of the decade, they ac-

---

1. Light travels along the surface of the sphere in this example, not off into space.

cumulated a catalog of dozens of distant type Ia supernovae observations, including measurements of supernovae that had exploded about seven billion years ago—roughly half of the age of the Universe. The astronomers, working in two groups, the Supernova Cosmology Project and the High-Z Supernova Search Team, hoped that their observations would help reveal the geometry of our Universe by measuring its rate of expansion.[2] If these distant supernova measurements were made with enough precision, the teams thought, they could determine at what rate the Universe's expansion rate was slowing down. If it was slowing down rapidly, then that would suggest that the Universe was closed, and the Universe would eventually cease expanding and collapse in on itself in a sort of Big Bang in reverse—a big crunch. If our Universe's expansion was found to be slowing down only mildly, then we would have learned that we live in an open space that would continue to expand forever.

When all of the supernovae observations were taken together, a trend began to emerge. The most distant of these supernovae consistently appeared to be less bright than had been anticipated. Against all expectations, these observations led to the conclusion that our Universe's expansion was not slowing down at all, but was accelerating.

The difference between predictions and results could not have been more stark. None of the three cosmological models I described earlier in this chapter—whether an open, closed, or flat Universe—predicted an accelerating rate of expansion. As it turns out, the answer to this problem lies with Einstein and his cosmological constant.

---

2. At the time, the BOOMERANG collaboration hadn't yet announced its determination that our Universe's geometry is flat.

• • •

Einstein, as I described in the last chapter, had originally introduced the cosmological constant into his equations in order to force the mathematics to provide a Universe that was static, neither expanding nor contracting. Although he later abandoned this extra term after Hubble's observations demonstrated that the Universe was in fact expanding, today it seems that perhaps he shouldn't have been so eager to throw away this extra piece of mathematics after all.

The way that the extra term in Einstein's equations affects the behavior of the Universe depends on its value. The value of the cosmological constant originally chosen by Einstein led to a static and unchanging Universe. But with a different value, it is also possible for such a term to cause the expansion of the Universe to accelerate over time. To their surprise, the Supernova Cosmology Project and High-Z Supernova Search Team had done much more than measure the expansion history of our Universe. They had discovered the existence of Einstein's cosmological constant! Nearly ninety years after the invention of the theory of general relativity, Einstein's extra term had been reintroduced into his cosmological equations. Although his original reasoning had been flawed, in a way, Einstein had finally been vindicated.

In 1998 the magazine *Science* declared this discovery to be the "Breakthrough of the Year." But what had been discovered? Einstein's cosmological constant is only a number in an equation. At least at first glance it is not obvious what such a term represents in nature. What is this strange "force" that brings about the acceleration of our Universe? In a way, it is nothing but space itself.

Physicists have coined many names for the energy contained in "empty" space, including phrases such as vacuum energy and zero-point energy. Cosmologists these days call it dark energy.

Assuming that dark energy exists in the same quantity in every piece of space and at all times, its presence would have precisely the effect of a cosmological constant. Dark energy does not act at all like the ordinary energy or matter in our Universe, however. Whereas ordinary energy or matter causes the Universe to be pulled together by the force of gravity, the presence of dark energy has a sort of counter-gravity effect, pushing the Universe apart.

Also, unlike the other components of our world, dark energy cannot be diluted. The expansion of the Universe reduces the density of matter and radiation. If the volume of the Universe doubles over a period of time as it expands, then the density of matter, including dark matter, is reduced by half. The density of dark energy, on the other hand, stays fixed. One cubic meter of space today has the same amount of dark energy that one cubic meter of space had a billion years ago or at any other time in our Universe's history.[3] This means that in the early Universe, when the density of matter was extraordinarily high, the relative quantity of dark energy would have been minuscule. As the Universe expanded and the density of matter was depleted, however, dark energy began to play an increasingly important role. Eventually, as more and more of our Universe was made up of dark energy, the expansion rate began to accelerate. Several billion years ago, the density of dark energy overtook the density of matter in our Universe to become the dominant component. Today, more than two-thirds of our Universe's energy is in this strange form. As time goes on, the density of matter in our world will become

---

3. This does not necessarily have to be the case, however. Models in which the density of dark energy varies with time have been proposed. In such scenarios, instead of Einstein's cosmological constant, the Universe's dark energy is something dynamic called quintessence. I will return to this possibility in the next chapter.

smaller and smaller. Our Universe's destiny, it now seems, is to expand faster and faster, until eventually the only thing that remains is dark energy.

· · ·

Combined with studies of the cosmic microwave background, the observations of distant supernovae provide a kind of inventory of our Universe. They tell us that the total density of matter and energy together is—to within 1 or 2 percent—equal to the critical density our Universe. Our Universe is either flat or very nearly so. Of this mass and energy, however, about 70 percent of it is not in the form of matter, but is dark energy. Of the remaining 30 percent, the majority is dark matter. A measly 4 percent of our Universe is in the form of atoms. Our visible world makes up less than one-twentieth of our Universe's total density.

Confidence in these conclusions has been strengthened by a wide range of observations and measurements. The amount of matter—both dark and visible—measured to be present in clusters of galaxies corresponds to only about a third of the critical density, not nearly enough to make our Universe flat without the contribution of dark energy. Measurements of the abundance of the light elements have been used to further reveal the quantity of atoms present in our Universe.[4] The WMAP satellite has measured the temperature fluctuations in the cosmic microwave background to even greater precision than its predecessors. Hundreds of distant supernovae have been observed and measured since the discovery of dark energy in 1998. Taken together, these data have

---

4. The abundances of the light elements are one of the primary pieces of information used to determine the 4 percent figure cited in the previous paragraph.

led to an overwhelming consensus in the community of cosmologists that we live in a Universe dominated by a combination of dark energy and dark matter.

New studies of distant supernovae, the cosmic microwave background, and other aspects of our Universe are also under way. Following the completion of the mission of the WMAP satellite, another experiment, called Planck, will set forth to map the cosmic background in even greater detail. A proposed project called SNAP—the Supernova Acceleration Probe—hopes to measure in detail the expansion history of our Universe by observing about 2,000 distant supernovae each year.

Just as the last decade has led to colossal advances in our understanding of our Universe, it is likely that the same will be true for the decade to follow. Just as the discovery of dark energy was not anticipated by many cosmologists, it is not difficult to imagine the next decade or two yielding other great, and unanticipated, discoveries. Although discoveries in science are rarely predicted, I would argue that it is very unlikely that we have seen the last great discovery of my lifetime in the field of cosmology. Another "discovery of the year" may be lurking around the corner. Perhaps even the discovery of a generation.

CHAPTER 10

AN UNLIKELY UNIVERSE?

No theoretical model, not even the most modern, such as supersymmetry or string theory, is able to explain the presence of this mysterious dark energy in the amount that our observations require.

—Antonio Riotto

The more I examine the Universe and study the details of its architecture, the more evidence I find that the Universe in some sense must have known we were coming.

—Freeman Dyson

The advances made in observational cosmology during the past few decades have been incredible. Astrophysicists not only have detected the cosmic microwave background, but have measured it in such detail that the geometry of our Universe has been determined to within a few percent precision—an impossible feat a generation before. In addition, the ever-increasing catalog of distant supernova observations has led to the discovery that dark energy comprises about 70 percent of our Universe's density. These findings are remarkable. Sadly, the attempts to develop a compelling theoretical explanation for these findings have been not nearly as successful.

It's fair to say that the theoretical physics community is, at least for the time being, entirely baffled when it comes to dark energy. It is not the fact that dark energy exists that is so confusing to us—that can easily be understood within the context of quantum theory. Instead, the thing that appears to be so inexplicable is the *quantity* of dark energy present in our Universe. For one thing, the density of dark energy is quite similar to the density of matter in our Universe, the former being roughly two to three times as great as the latter. Dark energy and matter are, as far as we understand them, completely unrelated phenomena. Finding out that they are present in roughly the same amounts is like finding out that the number of grains of sand on Pebble Beach is roughly equal to the number of spots on all of the Dalmatians throughout the history of the world. If that were true (and it is very likely not) it would either suggest that there is some deep underlying principle that relates grains of sand to spots on dogs, or it would simply be a remarkable—and incredibly unlikely—coincidence.

Although there are probably not many beachcombing firemen out there who lose sleep at night worrying about this kind of question, there are certainly many cosmologists who do. Does the numerical similarity between the amounts of dark energy and matter signify a deep connection between the two, or do we live in a remarkably—and I mean *remarkably*—unlikely Universe?

• • •

The constant emergence and destruction of virtual particles, as described by quantum physics, could be the key to the mystery. Because we believe that we know and understand those laws, it should in principle be possible to use them to calculate how much dark energy virtual particles generate. To perform such a calculation, however, all types of particles need to be taken into account, including those that have not yet been discovered. That limitation makes it impossible to calculate precisely how much dark energy these particles might produce. Nevertheless, particle physicists can still estimate how much dark energy should be generated by the fluctuations of quantum particles in the vacuum. When those calculations are carried out, a problem emerges. The universe they predict is not at all like our own.

In this kind of calculation, we are forced to make estimates for some unknown quantities, so you might expect that there would be some modest discrepancies. What is found, however, is that the estimates don't come even remotely close to the quantity of dark energy that is observed in our Universe. A simple version of the calculation, including only the known particles of the Standard Model, finds that there should be a whopping 10^{120} times more dark energy than the quantity we observe. That is a 1 with 120 zeros following it. Most people find it difficult to mentally distinguish large numbers like millions and billions and trillions. To me and nearly everyone else I know, 10^{120} is an incomprehensibly large number. Trying a simple thought experiment will demonstrate, if not how large that number is, at least how inconceivably large it is.

Consider the national debt of the United States. On February 1, 2006, it was roughly \$10 trillion.[1] That can be written as \$$10^{13}$.

---

1. According to the U.S. Bureau of the Public Debt's Web site, it was \$8,183,138,191,456.56.

Now imagine that there are 10^{13} nations with equivalent debt per planet, 10^{13} planets per solar system, 10^{13} solar systems per galaxy, 10^{13} galaxies per galaxy cluster, 10^{13} galaxy clusters per super-cluster, and 10^{13} superclusters per universe. The total indebtedness of all those nations on all those worlds would be only $\$10^{91}$. You would still need there to be 10^{13} universe clusters and 10^{13} clusters of universe clusters to get to $\$10^{117}$, which is still one thousand times smaller than $\$10^{120}$. You can see that the scale of the discrepancy between the amount of dark energy we observe and what the Standard Model predicts is beyond colossal. A trillion here and a trillion there, you might say, and pretty soon you're talking real money.

The difference between the expected quantity of dark energy and reality was known to be a problem even before any measurements of the dark energy density were made. If there were really as much dark energy in our world as the calculations would lead us to believe there should be, things would be very different. The relatively modest quantity of dark energy observed in our Universe caused the expansion of the Universe to begin accelerating a few billion years ago—the recent past, cosmologically speaking. If there were 10^{120} times more dark energy present, our Universe would not have begun accelerating so recently, but instead would have exploded at a colossal rate long before structures such as galaxies, stars, and planets had had a chance to form. As a result, the world would be nothing but vast emptiness. Even if the amount of dark energy were a mere ten or one hundred times more than the amount we observe, the matter that constitutes those structures could never have come together. It would seem that the estimates of how much dark energy should exist in our Universe must be flawed. This perplexing situation has come to be known as the cosmological constant problem.

Once again, the Standard Model needs to be saved from itself.

Because symmetries fixed earlier problems in theories, it is plausible to imagine that some as-of-yet-unknown symmetry might solve this problem. For example, a symmetry could exist that causes all of the quantum contributions to the dark energy in our Universe to exactly cancel each other. Perfectly unbroken supersymmetry could accomplish this, for example, with every contribution to the dark energy generated by a boson being exactly cancelled by the contribution of its superpartner fermion. We know, however, that if supersymmetry exists, it is broken. Broken supersymmetry, instead of removing dark energy entirely from our Universe, predicts an amount of dark energy roughly 10^{60} times greater than is observed. Furthermore, even if particle physicists can imagine how an unbroken symmetry might remove dark energy entirely, or how a broken symmetry could reduce it somewhat, they have a much more difficult time coming up with a symmetry that would reduce the amount of dark energy to the observed—comparatively tiny, but nonzero—quantity. Symmetry arguments have not provided us with a satisfactory explanation.

Getting no help from symmetry, many scientists have explored far more radical solutions to the cosmological constant problem. It is not at all clear where these lines of thought will lead us. There is little doubt, however, that whatever solution to the problem is eventually found, it will force us to dramatically change the way we think about our Universe.

• • •

When faced with a problem that has no apparent solution, one approach is to wonder whether you are asking the wrong question. Perhaps the solution to the cosmological constant problem is that the dark energy we observe in our Universe isn't a cosmological constant at all. But what is the alternative to the dark energy being Einstein's cosmological constant? Well, we can imagine instead that dark energy might not be constant over space and time, but might be something whose energy density could vary between points in space, or over time. We might even go so far as to imagine something dynamic, such that over time its density gradually came to approximate that of matter. In this way, we might hope to explain why these two seemingly unrelated things—matter and dark energy—appear in such similar quantities in our Universe. The most extensively studied example of this kind of dynamic dark energy is called quintessence.

In theories of quintessence, dark energy is generated by a new type of ultralight particle whose effect on the energy density of space gradually changes with time. In some of these models, quintessence interacts in such a way that causes its density to approach the density of the ordinary (that is, not dark) energy in the early Universe. In the very early Universe, that ordinary energy was predominantly in the form of radiation.[2] As time passed, however, the quantity of radiation relative to matter decreased, and eventually matter became the greatest constituent part of our Universe. It would have been at that stage that the quintessence began to act like a cosmological constant form of dark energy, with its density nearly fixed, varying only slightly with time. Many cosmologists find this model attractive because the amount of dark energy that would be generated in this way is not the enormous 10^{60} or 10^{120} times the quantity we observe. Instead, in a Universe in which quintessence and ordinary matter and energy are related, a quantity of dark energy roughly equal to that observed could be generated.

But is quintessence a plausible substitute for a cosmological

---

2. By radiation here, I mean particles with most of their energy in the form of motion (kinetic energy) rather than mass. By matter, I mean the opposite.

constant form of dark energy? If our Universe's dark energy is not the result of a cosmological constant but instead is something dynamic—like quintessence—then the expansion history of our Universe could be quite different (figure 10.1). Quintessence, like a cosmological constant, can cause the expansion rate of the Universe to accelerate, but not necessarily at exactly the same time or in exactly the same way. Very precise measurements of large numbers of distant supernovae carried out in the future may enable us to test those different predictions. Current supernova observations, however, do not yet allow for such conclusions to be drawn. Although the expansion history as we can presently measure it appears to be what a cosmological constant would generate, the margin of error likely allows for a range of possible quintessence models. For the time being, supernovae can neither confirm nor exclude the possibility of quintessence.

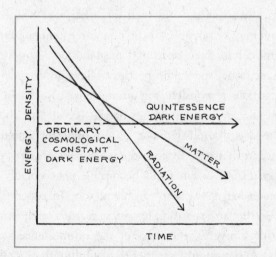

FIGURE 10.1. If dark energy takes the form of quintessence, it would have had an energy density in the very early Universe similar to that of radiation. Around the time that matter became more plentiful than radiation, the quintessence would have begun acting like a cosmological constant.

Despite the appeal of quintessence as a solution to the cosmological constant problem, and although quintessence theory appears to be consistent with supernova observations, serious theoretical problems hamper the idea. One problem comes from the fact that the ultralight particle needed in such models is an example of something called a scalar field.

Scalar fields are a type of boson, the class of particles responsible for the forces found in nature. Although we have not yet observed a fundamental scalar field in nature, most particle physicists believe they are likely to exist—the yet-undiscovered Higgs boson, for example, is a scalar field. As I've said, this quintessence particle would have to be extremely light. That creates a problem, because the lighter a boson is, the longer will be the range of its corresponding force. A scalar field light enough to generate our Universe's dark energy in the form of quintessence would correspond to a new force that would act over very long distances, even cosmologically speaking. No such force has ever been detected. Although it may be possible for the new force generated in a quintessence model to have remained hidden, such a model would have to have rather special properties to have gone undetected.

The particle responsible for quintessence also has a problem similar to what dogged the Higgs boson. Recall that the interactions of the Higgs boson with other kinds of particles tend to make it heavier, and that only by introducing supersymmetry could we prevent the Higgs boson from becoming unacceptably massive. Quintessence has a "weight" problem, too. In order for such a model to predict an expansion history consistent with the observations of our Universe, the mass of the quintessence scalar field would have to be roughly a trillionth of a billionth of a billionth of a billionth of the mass of the electron. It is next to impossible to imagine how a scalar field could stay so unbelievably light.

These serious theoretical obstacles notwithstanding, cosmologists continue to explore the possibilities associated with quintessence. The fact that such models continue to be studied is a testament to just how desperate cosmologists are to understand the nature of dark energy. Quintessence is not the most desperate solution on offer, however. Another solution is probably the most controversial idea in all of modern physics.

• • •

One of the central tenets of cosmology is that there is nothing unusual about the corner of the Universe we find ourselves in. This is often called the Copernican Principle, after Copernicus's realization that Earth does not sit at the center of the solar system or Universe. Just as our planet is not unique, neither is our solar system or galaxy—in the grand scheme, wherever you look, the Universe is pretty much the same.

Our Universe is enormous, and the overwhelming majority of it is nearly empty, with next to no matter. Nevertheless, in all the vast volume of our Universe, we happen to live in one of the rare corners of space that contain a lot of matter in the form of a planet. Do we think of this as unlikely or serendipitous? Of course not. We live in a matter-rich region of our Universe because we are matter-rich beings who need matter to live. Raw materials—chemical compounds—are needed in sufficient quantities to supply life forms with their required energy and building materials. Life also needs some degree of protection to develop and survive. Features of our planet serve as bulwarks against everything from cosmic radiation and meteor collisions to other sudden environmental changes. Life, at least life such as we have on Earth, cannot emerge from just anywhere. We should not be surprised to find ourselves on the surface of a planet from which life could emerge. If you stop to think about it, it could not be otherwise.

Some scientists have applied this logic to the dark energy problem, too. As I said before, if dark energy existed in the quantity estimated by the relevant particle physics calculations—10^{60} or 10^{120} times more than we observe—then the Universe would have expanded so quickly that structures like galaxies, stars, and planets would never have had a chance to form. Certainly no life could have come out of such a Universe. Because we exist in a world in which life has developed, our Universe cannot be one with such a large amount of dark energy. Even if the most likely quantity of dark energy for a universe to contain is roughly the value estimated by particle physicists, we must find ourselves living in a universe like ours, with much less dark energy, or else we wouldn't exist.

Dark energy isn't the only feature of our Universe that has compelled scientists to think about their world in this way. Take, for example, the strengths of the forces of nature. If the strong nuclear force were a mere 4 percent stronger than it is, pairs of protons would bind together into diprotons, which would cause stars to burn their nuclear fuel at a rate a billion billion times faster than they do, essentially causing all stars to instantly explode. If the strong nuclear force were 50 percent weaker than it is, or if the electromagnetic force were about twice as strong as it is, carbon atoms would disintegrate—so much for carbon-based life. The more closely one looks at the conditions present in our Universe, the more it appears that even small changes in its structure would be catastrophic for the prospects of life.

The counterargument to this reasoning is that although the Universe would likely not be capable of supporting life as we know it if it were different, it still might be able to generate and support other forms of life. From this point of view, the fact that our Universe contains the kind of life that we know is not particularly profound. It is not obvious how strong this counterargument is, however. Although it is clear that it is valid in some cases, there are some kinds of universes one could imagine that could not support life of any kind. A universe capable of supporting life must contain matter that can undergo an array of complex interactions. An empty or non-interacting universe cannot lead to life. Similarly, a life-supporting universe must, to some degree, be stable. There are some features that a universe must have if it is to support life of any kind.

• • •

These arguments, and others like them, are based on a philosophical concept known as the anthropic principle. This idea, introduced in 1974 by the theoretical physicist Brandon Carter, comes in many forms. Succinctly, the anthropic principle holds that any physical theory we formulate must allow for the possibility of life. Put more prosaically, the principle is a statement that for us to exist, the Universe must have a set of characteristics that allow for life to develop within it, because we have developed within it.

To the theologically minded, the anthropic principle has sometimes been seen as an argument for a creator who deliberately designed our Universe with the intention of it containing life.[3] This is not by any means the only plausible interpretation,

---

3. At the time at which I am writing this book, a public debate has been taking place in the United States about the theory of intelligent design. I do not want what I am writing in this chapter to be misunderstood as support for this approach in science in any way. The possibility that our Universe was designed by an intelligent being has no way of being tested experimentally, and therefore is not a scientific hypothesis. This does not make it wrong, but it does make it nonscientific.

however. In order to be asking these kinds of questions, we—a form of life—must be present in the Universe. Thus the anthropic principle demonstrates itself without any need for introducing a "creator." Furthermore, if there were not only our Universe but many universes—each with different properties—in existence, then most of those universes would be devoid of life simply because they lacked the rather specific properties required to support such complexity. Given enough different universes, however, life would exist in some number of them. In these rare, life-friendly universes, the forces and types of matter that exist would lead to a rich and complex set of interactions—such as those possessed by the atoms and molecules of our world, for example. Additionally, the quantity of dark energy present in these life-friendly universes would be far below the enormous values estimated by particle physics calculations. It would be in one of these rare universes in which we live.

The anthropic principle has enjoyed some popularity among scientists recently, buoyed in part by recent developments in string theory. Some string theorists have been "counting" the number of possible universes that can exist in their theory. That might seem like an impossible task. After all, if we imagine a world that had a slight change to just one constant of nature—the strength of the electromagnetic force, for example—it would be different from ours. It seems, then, that an infinite number of possible universes could exist. Such reasoning doesn't necessarily apply in string theory, however. Instead of being entered into a theory by hand, in string theory the constants of nature are derived from the theory itself. As a result, string theory appears to predict a finite, although enormous, number of possible universes, or string vacua, as they are called. The current estimate of the number of possible string vacua is something like 10^{500}. Yes, that is a 1 with 500 zeros after it. My recommendation is to not even try to intuitively understand the colossal size of this number. I know I can't.[4]

If string theory does describe the nature of our Universe, it would be strange that it could predict so many possible universes if only ours existed. With this in mind, many string theorists have come to adopt the idea that numerous universes are likely to exist. It is under such conditions that the anthropic principle thrives.

• • •

To say that anthropic reasoning is controversial in science is quite an understatement. Rarely are the opponents or the advocates of the anthropic principle the types who agree to disagree; rather, they are more often the shouting/clawing/biting types. If nothing else, the vitriol can make for interesting "discussions" at cosmology conferences.

Many of the objections raised against anthropic reasoning have correctly pointed out that some scientists misuse and overextend the anthropic principle beyond its logical limits. Most scientists are, after all, not skilled philosophers. Even when an anthropic approach is taken within its logical limits and without such flaws, its use still often draws strong criticism.

The reason most often given for such opposition is that some scientists see the use of the anthropic principle as essentially giving up on trying to explain our world and just adopting a "that's just the way it is" philosophy. The advancement of science has always been made possible through continuously refining our observations and experiments until at some point they no longer

---

4. Other recent estimates of this quantity have found numbers ranging from about 10^{250} to 10^{1000}. For our purposes, the precise size of this number is of little importance. Just think of it as a hell of a big number!

match the predictions of our theories. When this happens, we develop a new and more complete theory that describes all of the data and observations. Then we test that new theory until we find out under what circumstances it fails, and so on. Over time, this process leads to increasingly complete and accurate theories. The opponents of anthropic reasoning argue that once we rely on the anthropic principle to explain why something is the way it is— why dark energy exists in such a small quantity, for example—this process ceases. If we were to say that dark energy exists in the amount it does because otherwise we would not exist, then we no longer have any reason to think we could ever discover a theory that explains *why* it exists in such a small quantity. In an anthropic framework, such a theory does not need to exist.

On the other hand, it is possible that a very large number of universes do exist and that our world is the way it is only due to a series of anthropic coincidences. In this case, there might be no fundamental theory that explains why the quantity of dark energy we observe is what it is. The opponents of the use of anthropic principle in cosmology generally agree that this is a possibility. Essential to them, however, is the fact that *we do not know with certainty that it is true.* As long as there might exist a theory that explains the nature of our world, we should continue to search for it. A surgeon doesn't stop an operation because the patient might not survive. Likewise, physicists should continue searching for greater and more complete theories of nature until we are absolutely sure that no such theory exists. In other words, we should continue searching forever.

· · ·

Not surprisingly, not all scientists agree with these anti-anthropic arguments. Among the proponents of anthropic reasoning are prominent physicists, including Nobel laureates. Instead of focusing on the limits of anthropic reasoning, these advocates see incredible possibilities for discovery. If it were possible to demonstrate that our Universe had the specific properties that were predicted by a theory of a set of universes, perhaps we could in this way actually learn of the existence of universes other than our own. Science might not be absolutely limited to the study of our own Universe as has often been assumed. Science might someday unveil a greater Multiverse.

Imagine that you had a complete theory that not only described our Universe, but predicted the set of all possible universes—a true Theory of Everything. Such a theory might even describe how a universe could come into existence. With such a theory in hand, it is at least hypothetically possible that it could be used to calculate the probability of a given universe having a given set of properties. Combining this knowledge with what properties are necessary for a universe to support life, it would then be possible to calculate the probability of a universe being a habitable universe. In this way, we could potentially determine the probability of our world being the way it is.

Although we don't yet have such a complete theory, some future generation of physicists might indeed develop one. This theory might be some more fully realized version of string theory, or might come in some other, yet unconceived, form. Whatever this theory might be, imagine that it predicts that 99.99999 percent of inhabitable universes contain some specific set of properties. If our Universe had this specific set of properties—the types of particles that exist, the values of the various constants of nature, the geometry of the space-time, and so on—it would provide evidence in favor of the theory. If, in contrast, a competing theory found that only 0.0000001 percent of all inhabitable universes resembled ours, we would be compelled to discard that theory.

Although both theories predict that universes like ours do exist, the first theory predicts that such universes are very common, while according to the second they are exceedingly rare. In this way, it might be possible that different theories describing the entire Multiverse might be experimentally distinguished. A theory extending beyond our world alone, extending to the Multiverse, could become science.

Before any such study could ever be conducted, however, a complete theory of all universes and their relationship to each other must be found. But how can distinct universes be "related"? And for that matter, where does a universe come from—from what does it emerge? Can one universe be "born" of another?

CHAPTER 11

COSMIC OFFSPRING

This idea of multiple Universes, or multiple realities, has been around for centuries. This scientific justification for it, however, is new.

—Paul Davies

If inflation happens once, an infinite number of universes are produced.

—Alan Guth

By the late 1970s the Standard Model of particle physics had become fairly well established. String theory, on the other hand, had not yet become a widely studied subject. During that period of time, many scientists in the particle physics community were working toward the goal of developing a grand unified theory—a GUT, for short. Just as Einstein had sought to unify electromagnetism and gravity, these physicists sought to develop a theory that combined each of the strong, weak, and electromagnetic forces. Such a theory, it was hoped, would incorporate all of the matter and forces of the Standard Model within a simple and elegant framework. The first modern grand unified theory was proposed in 1974, and at the time it appeared likely that such a theory was soon to be confirmed.[1]

Such optimism proved to be unfounded. Many of the proposed grand unified theories predicted that protons should decay, although only very rarely.[2] With this prediction in mind, experiments were developed to search for signs of this occurring.[3] When these experiments detected no sign of proton decay, many of the simplest GUTs were disproven. Despite that failure, many other models of grand unification remain experimentally viable and continue to be studied today.

But there was another problem. According to GUT models, a transition took place early in the history of the Universe. In this transition, the three forces of the Standard Model—the electromagnetic, strong, and weak forces—emerged from a single grand unified force. As a consequence of this process, an enormous number of strange objects called magnetic monopoles would be generated. If so, it was thought, these magnetic monopoles should still be scattered throughout our Universe today.

A magnetic monopole is a particle with a net magnetic

---

1. A grand unified theory or GUT differs from a Theory of Everything in that it does not incorporate gravity.

2. You might worry that proton decay would have caused the atoms and molecules in our Universe to have disintegrated. The predicted rate of proton decay in these theories is so low, however, that the overwhelming majority of protons created in the Big Bang would still be around today.

3. In chapter 4, I discussed the Super-K experiment, which was built to look for decaying protons but ended up primarily revealing aspects of the nature of neutrinos.

charge—essentially, it's a magnet with only a "north" or "south" end. If that seems like a ludicrous idea, most likely you've never even considered the possibility of magnetic charge. You're in good company. More than one hundred years ago, the physicist Pierre Curie noted that James Clerk Maxwell's basic equations describing the properties of electricity and magnetism had a special property. Maxwell's equations are nearly symmetric with respect to electricity and magnetism. The symmetry is broken in only one way: electric charge exists, but magnetic charge does not. An electron, for example, has a net quantity of electric charge. This makes an electron an electric monopole. A magnetic monopole is an object with a net magnetic charge—a "north" without a "south," or vice versa. So far, no magnetic monopole has ever been found to exist in nature.

But what about magnets? Don't they have net magnetic charge? The answer is no. To see why, picture a simple bar magnet. Such a magnet has both a "north" and a "south" end, or pole. If you were to experiment with a pair of bar magnets together, you would find that when you bring two north poles or two south poles together, they repel each other. If, on the other hand, you bring a north pole and a south pole together, they attract one another. Each end has an equal but opposite charge. This basic bar magnet, instead of being a magnetic monopole, has both north and south pole—a magnetic dipole.

But now let's say we wanted to have a magnet with only a north pole, without the corresponding south end, a magnetic monopole. To do this, we might first try breaking a bar magnet's north pole off from its south. This, however, leaves us with two magnets, each with its own north and south pole (figure 11.1). We have never observed a magnetic north pole without a corresponding south pole. A positive magnetic charge, which we'll call north, has always been found to be accompanied by an equal negative

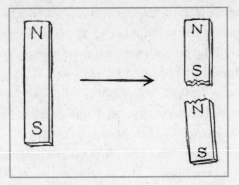

FIGURE 11.1. When a bar magnet is broken in two, we are not left with isolated north and south poles. Instead, two bar magnets are formed, each with its own north and south pole.

magnetic charge, which we'll call south. An object with a net magnetic charge has never been seen in our Universe.

Although Maxwell's equations assume that magnetic monopoles do not exist, we do not know for certain that this is true. The possibility that magnetic monopoles might exist and just haven't been seen yet is even more interesting within the context of quantum theory. The mathematics of quantum physics can be used to show that if at least one magnetic monopole did exist, then electric charge must exist only in specific, quantized, amounts. And, of course, in our Universe electric charge does exist in such quantized amounts. Every object in our world has a quantity of electric charge exactly equal to that of an electron multiplied by an integer (any whole number, positive or negative). For convenience, we define an electron's charge to be −1; thus, the charge of a proton is +1, and the charge of a photon is 0. All electric charge comes in packets of that fundamental quantity, −1. An object with 1.2343 times the electron's charge, for example, cannot exist if even a single magnetic monopole is present in our Universe.[4]

The symmetry between electricity and magnetism nearly found in Maxwell's equations is restored in grand unified theories. The only reason that we don't observe and haven't created magnetic monopoles in experiments is that these theories also predict that magnetic monopoles are very heavy, containing more energy than a million billion protons—far surpassing what we can generate in particle accelerators. The Big Bang, however, had more than enough energy. In the first moments following the Big Bang, the Universe was hot enough to create magnetic monopoles in enormous quantities. And once created, magnetic monopoles are very difficult to destroy. An isolated magnetic monopole cannot decay into any lighter particles because the total amount of magnetic charge in any interaction must be conserved, and no lighter magnetically charged particles exist.

The first estimates of how many magnetic monopoles had been created in the Big Bang, and how many should remain today, were made in the 1970s. Although the answer to this question varies somewhat from theory to theory, all GUTs predict that magnetic monopoles should be roughly as plentiful as protons in our Universe. Each monopole, however, would contain at least a million billion times more energy than a proton, implying a combined energy density much, much greater than is observed.

---

4. At first glance, quarks may seem to be an exception to this rule. Individually, quarks contain electric charge in a quantity of either $1/3$ or $2/3$ that of an electron. Quarks cannot exist individually in nature, however, and can only be found in groups of two or more bound together. In any such bound group, their electric charge always adds up to an integer multiple of the electron's charge, thus not violating this rule.

Obviously, monopoles in such enormous quantities cannot exist. GUTs were in trouble.

• • •

At around the same time that particle physicists were wrestling with the problem of the over-abundance of magnetic monopoles, many cosmologists were asking why the geometry of the Universe was flat, or at least so close to it. When the density of our Universe was determined to be the value corresponding to a flat universe, or at least a universe close to flat, this appeared very strange to many cosmologists. As time passes and a universe expands, if its density is not exactly the critical density (the value corresponding to a flat universe), it steadily becomes more curved and less flat. If in the first seconds after the Big Bang the Universe had a density that was even slightly more or slightly less than the critical density, then by today the Universe would be very highly curved. In fact, in order for the density of our Universe to be within a factor of ten of the critical density today, the density at one second after the Big Bang would have to be fixed to the exact value of the critical density to within one part in 10^{60}. Finding our Universe to be flat is like balancing a needle on a table and finding that it hadn't fallen over a billion years later.

Furthermore, the homogeneity of the cosmic microwave background also posed a problem to cosmologists. In an expanding Universe, there are limits to how far anything—including information—can travel. As the Universe expands, all points in space steadily move farther apart from each other. Two points can be so distant from one another that the space between them grows at a rate faster than the speed of light, making communication between the two points impossible. In effect, each point in space is surrounded by a horizon. It is impossible to send information beyond this horizon, or to receive information from outside of it. The horizon acts as an impenetrable shroud, billions of light years away, concealing everything beyond it.

We can picture our horizon as an enormous sphere surrounding Earth. Every other point in space has its own horizon as well. Two points on opposite sides of our horizon fall outside of each other's horizons.

Two such points that are outside of each other's horizon are said to be causally disconnected—unable to interact with or pass information between each other. Because two causally disconnected regions cannot affect each other in any way, we would expect that such separated parts of our Universe would have developed somewhat differently. For example, it was expected that different parts of the cosmic microwave background would be measured to have different temperatures. As we've seen, however, the variation in the background temperature in the sky is minuscule. Somehow, these vastly separated regions of our Universe were in contact with each other—they "knew" about each other's temperature. Somehow the most distant regions of our Universe had been causally connected to each other.

• • •

These three puzzles—the monopole, flatness, and horizon problems—confounded cosmologists and particle physicists. By the end of the 1970s, however, a solution to all three of these problems had been proposed. This solution came in the form of a theory now known as inflation, which was first developed by Alexei Starobinsky, a Soviet cosmologist. According to this theory, our Universe underwent a very brief period of incredible, ultra-fast expansion in its early history. Although this proposal might appear wildly speculative or ad hoc, much of the work being conducted at the time on grand unified theories showed that such a scenario was plausible. Many of these theories predicted the existence of scalar fields—precisely the ingredient needed to bring about the sudden burst of growth and expansion that we now call inflation.[5]

Once a period of inflation is initiated, the size of the Universe begins to double nearly once every 10^{-32} seconds. It is easy to see that at this rate, it would take almost no time at all for a universe to grow from an extremely compact and densely filled state into an enormous volume of space. In most models the period of inflation lasts just 10^{-30} seconds, after which the Universe's expansion rate returns to a steady and much slower progression. Nevertheless, during that brief time, inflation theory predicts, the Universe grew larger by at least a factor of 10^{25}, and possibly much more.

As the Universe inflated, the number of monopoles present in a given volume became diluted by the enormous increase in the volume of space. If you take a small balloon, put a collection of dots on its surface with a marker, and then inflate it to a much larger size, the number of dots per square inch of the balloon's surface becomes much smaller. Similarly, even if an enormous number of monopoles existed in the early Universe, their density is reduced so dramatically during inflation that we would most likely never observe one in our Universe today.

Regarding the flatness problem, in contrast to a steadily expanding universe, inflation tends to bring the density of energy and matter in a universe toward the value corresponding to a flat universe. Picture again an inflating balloon. If it contains only a

---

5. You might recall that models of dark energy in the form of quintessence also contain scalar fields. The Higgs boson I described earlier is also an example of a scalar field.

little air inside of it, its surface will be considerably curved—a closed space, to use the language of general relativity. As the balloon is inflated, however, this curvature gradually decreases, and the surface becomes closer and closer to flat. When you stand on Earth, its surface appears, on average, to be very nearly flat. If instead you were standing on a basketball, the curvature would be much more evident. Therefore, if inflation did take place in the first moments following the Big Bang, it is not a surprise that we find ourselves living in a universe that appears flat. Instead of being incredibly unlikely, a flat universe is an unavoidable consequence of the inflation theory.

Inflation solves the horizon problem as well. Remember that the limitations of a horizon come from the fact that information cannot travel faster than the speed of light. There is nothing in Einstein's theory, however, that prevents space itself from expanding at a faster rate. This is precisely what happens during inflation.

Picture two objects in the Universe separated by a short distance just before the period of inflation begins. Suddenly inflation starts and a tiny fraction of a second later ends, and the Universe is much, much larger. These two particles, only a short distance apart a moment ago, are now nowhere near each other. This is not because either of them moved, but because the space between them expanded. Inflation suddenly moves everything away from everything else at a rate far beyond the speed of light, and without violating the laws of relativity.

After inflation ends, it might appear that our two hypothetical particles would be outside of each other's horizons. The fact is, however, that an instant ago they were in causal contact with each other. For this reason, the horizon of each particle is actually much larger than it would appear. Likewise, regions of the cosmic microwave background in opposite directions of the sky were, due to inflation, more recently causally connected than they appear to have been. Through the process of inflation, opposite ends of our visible Universe can appear to have been speaking to one another.

. . .

When he developed the first theory of inflation, Starobinsky was working at the Landau Institute of Theoretical Physics in Moscow. In the 1970s very little communication was possible between Soviet and western scientists.[6] Although Starobinsky's model became quite famous among Soviet cosmologists, it remained unknown in the West. In 1981 Alan Guth, unaware of Starobinsky's work, invented an inflationary model of his own in the United States. Guth's work became well known in western cosmological circles, and although his first model had some technical problems, Guth was successful in persuading many scientists of the merits and beauty of inflation theory.

Later that year, cosmologists from the two worlds of the Cold War era met together at a conference in Moscow. The new theory—or theories—of inflation were exciting and controversial topics. Among other prominent scientists attending the meeting, Stephen Hawking argued that no model of inflation could possibly work. Later in the meeting, Andrei Linde, a Russian cosmologist, presented his own theory of inflation that avoided the problems of Guth's theory. Linde's model ultimately persuaded the opponents of inflation, including Hawking, that it was possi-

6. I will resist the temptation to make a joke about causal disconnectedness here.

ble. Within a year's time, the theory of inflation had become a firmly established element of modern cosmology.

• • •

When the Universe suddenly inflated 14 billion years ago, it became very big, very fast. This kind of "bigness" is difficult to overstate. In fact, it might be impossible. Earlier in this chapter, I said that the period of inflation comes to an end very shortly after it begins. Technically speaking, that is only partially true. Inflation does tend to stop quickly almost everywhere, but not quite everywhere. Small fractions of the total volume can continue to expand rapidly even after the period of inflation has ended elsewhere. That is not as inconsequential as it might seem. A very small amount of space that continues to inflate becomes a very large amount of space very quickly. When most of that new large amount of space ceases to inflate, tiny fractions of its volume keep going, each becoming enormous. In this way inflation never really ends. Although it appears to end to us, as it would to most of the other hypothetical observers elsewhere in the Universe, some parts of space always continue to inflate. Inflation, it appears, is eternal (figure 11.2).

As we've seen, inflation causes space to expand into many causally disconnected regions. If inflation happens not only briefly, but eternally, then a practically infinite number of these regions are formed, each unable to interact with the others. It is possible that some of these other regions might even have different laws of physics than those we know. These different spaces are, in every practical sense, different universes.

I opened this chapter with a quote from Alan Guth: "If inflation happens once, an infinite number of universes are produced."

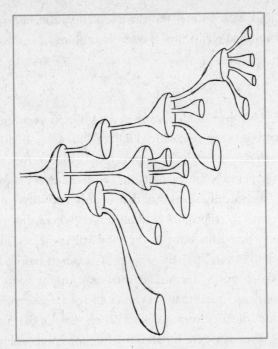

FIGURE 11.2. The mathematics of inflation theory suggest that when a universe inflates, some small regions of it are likely to continue inflating, even after the rest has stopped. These regions go on to inflate into very large volumes of space, some small parts of which will continue to inflate. This process never stops, and continues to bring forth more and more universes without end.

This appears to be an inescapable conclusion. Virtually all models of inflation that have been proposed continue to produce more and more universes as time goes on, never stopping.[7] If a period of

---

7. The argument for this is statistical. The probability of a point in space ceasing to inflate quickly is high (nearly one), but there is always a small but finite chance of a region continuing to inflate further. Imagine a set of decaying particles with a half-life of one second, but each of which turn into one million particles if it survives ten seconds. Most of the particles would quickly decay, but those that didn't would turn into many particles. Ultimately, this leads to an infinite number of particles. The same goes for universes.

inflation took place in our Universe's history, it would seem that we must conclude that ours is only one of a practically infinite number of universes.

• • •

And now, after a considerable detour, we finally return to the puzzle of dark energy. The observation that dark energy exists in our Universe in what appears to be a completely unlikely quantity—the cosmological constant problem—was the main theme of the previous chapter of this book. Inflation may provide a solution to the problem. As baby universes become parents themselves, inflation quickly leads to a practically infinite number of universes.

Of these many, many worlds, the vast majority will contain roughly 10^{60} or 10^{120} times more dark energy than the amount we find in our Universe, which, as we saw in the previous chapter, is what our particle physics calculations tell us to expect. These universes will all expand extremely rapidly and will never contain life of any kind. A fraction of the many universes, however, will have a little more or a little less dark energy—but still will not contain life. Furthermore, a very, very, very small fraction of these universes will contain much, much, much less dark energy. Some of these universes will contain so little dark energy that it would be billions of years before the effects of their dark energy would become significant. In such worlds, space expands slowly enough to allow for the formation of structures such as galaxies, stars, and planets. It is in one of these rare worlds that life might be able to form. It is in one of these worlds that we must invariably find ourselves living.

I will conclude this chapter with a word of caution. Although all of the ideas I have presented are within the mainstream of modern cosmology, it is far too early to know whether this is in fact the way our Universe—or universes—work. Inflation is a beautiful and incredibly successful theory. I am reasonably confident that it will, in some form, be further experimentally confirmed and understood in much greater detail in the coming years and decades. Furthermore, virtually all models of inflation that have been studied thus far suggest that once inflation begins, it will continue in some part of space, creating more and more universes forever. From our current level of understanding, eternal inflation seems like an inescapable conclusion. And with eternal inflation come a practically infinite number of different universes, the existence of which allows us to easily explain away the puzzle of why we find dark energy in our world in such a strange and unexpected quantity.

Still, with these things being said, I must admit that we do not yet understand enough about inflation to say that we are certain. It would be amazing to determine scientifically that other universes did, in fact, exist. It would be incredible to understand that our world came forth through the process of inflation as the offspring of a network of parent and baby universes. It is awesome to know that the first steps toward developing this idea were put forth within my own lifetime. Much work remains to be done, however. It is the task of this generation of physicists to explore more thoroughly this possibility and to confirm or reject their conclusions through the process of solid experimental testing. It is the task of this generation to find out if these ideas are right.

CHAPTER 12

THE SKEPTICS

If you would be a real seeker after truth, it is necessary that at least once in your life you doubt, as far as possible, all things.

—René Descartes

The trouble with the world is that the stupid are cocksure and the intelligent are full of doubt.

—Bertrand Russell

While living in Oxford, in addition to working on research I occasionally took on some teaching. Besides teaching the tutorials that are the foundation of the Oxford/Cambridge system, I also developed and taught a course that I named "The Weird and Wondrous Universe" for Oxford's department of continuing education.

Courses in the department of continuing education were not

filled with the typical Oxford students. These classes were open to the general public, and instead of being filled with twenty-year-old proto-academics and future bankers with proper-sounding names and smoking jackets, they were taken by people of a wide range of ages and backgrounds. The only thing they seemed to have in common was their sheer enthusiasm for learning (in this case) physics and cosmology. After all, they weren't there to get a degree or to further their career. They were there because their curiosity compelled them.

During each week's lecture, I was continually bombarded with questions relevant to the night's topic (although some were more relevant than others): What happens to something that falls into a black hole? If neutrinos can pass through Earth, how do we know they are there? What would happen if a nearby star exploded as a supernova? If protons are made up of quarks, are the quarks made up of something smaller? I loved these questions. Only a truly interested student would ask them—or even come up with them.

One student in the class had more questions than any other. This student, a retired engineer named John,[1] seldom let more than ten or fifteen minutes pass between his inquiries. John's questions and comments differed somewhat from those that other students would ask. He rarely accepted the things I would say as fact. In one of my first lectures, I introduced Einstein's relativity. I talked about how gravitation could be thought of as something that is not a force at all, but that is the consequence of the curvature of space-time. John's voice cut through the room, passionate as always: "But surely this must be wrong!" I could convince John that space-time was a useful graphical concept, but not necessarily that it represented the real world. A declaration that something

---

1. At "John's" request, I am not using his real name.

was true was not sufficient for John. I quickly learned that he would require a very careful argument to persuade.

I don't believe that John's reaction to these ideas arose from any doubts he might have had regarding my qualifications as a physicist. It seemed to me that I could have had two Nobel Prizes in my pocket and it would have been all the same to him. No, John was—and is—the kind of person whose instincts tell him to doubt an assertion more often than they tell him to accept something without solid proof. John is a skeptic.

• • •

Skepticism is one of the most important qualities a scientist can have. Without it, the orthodox perspective of a scientific generation will risk becoming fixed in place, without any possibility of advancement. If Einstein had not been skeptical of Newton's law of gravitation, general relativity would have had to wait for someone else to discover it. If Charles Darwin had not been skeptical that Earth and all of its forms of life were created, along with the rest of the cosmos, all at the same time—perhaps over a six-day span several thousand years ago—our understanding of evolution through the process of natural selection would not have come about when it did. If the great geometers of long ago had not been skeptics, perhaps we would still believe today that Earth is flat. The same is true in science today. Modern science has produced many wonderfully successful theories that agree remarkably well with all of the tests we have applied to them—the general theory of relativity and quantum theory being two excellent examples.

Nevertheless, as we've seen, those theories are not perfect and they can be improved. The overthrow of successful theories that were refined or replaced by more complete theories is what makes up most of the history of science. But even as a theory is replaced, it is important to recognize that this does not make the supplanted theory wrong. Newton's law of gravitation works very well under most conditions, so I don't think it's fair to call Newton's theory wrong. Einstein's general theory of relativity, however, includes everything Newton's theory does and it works in situations under which Newton's theory fails: in very strong gravitational fields or at very high speeds. Everyone agrees that Einstein's theory is better, meaning more useful and more widely applicable, than Newton's. Rather than saying one is right and another wrong, I think of them as being more and less complete.

That distinction pertains to all branches of science, including cosmology. The vast majority of modern cosmologists are convinced that around 14 billion years ago our Universe was in an ultra-hot state that expanded over time to eventually become the world we see before us today—the Big Bang. In fact, I don't believe that I have ever met a cosmologist who disagreed with this assessment. Furthermore, I think it is highly unlikely that future generations of cosmologists will have cause to replace or drastically revise this basic picture of our Universe's history. I also think that some of the details included in this description are likely to be revised or replaced in the future.

Certainly, if the past is any guide, then that is bound to be true. More than twenty years ago, the idea that a period of inflation took place in the early history of our Universe was introduced and accepted by most of the scientific community. This modification to the traditional picture of a more steadily expanding Universe did not overthrow the well-established theory of the Big Bang. It did, however, revise it. Other new elements will likely be introduced over time into future versions of the Big Bang theory. This kind of tweaking leads gradually to a more complete theory that describes more aspects of our world more fully.

What other new elements will be added to our picture of the Universe remains to be seen. Might we find that the next generation of cosmologists will hold different conclusions about how the first structures formed in our Universe? Might they argue that inflation occurred in a very different way than we have thus far imagined? Might they become convinced that either dark matter or dark energy—or both—does not exist? Whatever they find, those discoveries will be driven by the questions of the skeptics.

• • •

Throughout this book, I have documented a great deal of compelling evidence in favor of the existence of dark matter and, to a somewhat lesser extent, dark energy. We see the effects of those elusive substances in many ways, each contributing to our confidence in their existence. The presence of dark matter can be inferred from the rotation speeds of stars around the centers of galaxies, or in their imprint on the cosmic microwave background. Massive mini-galaxies called dwarf galaxies appear to be made up almost entirely of dark matter. Observations of large clusters of galaxies indicate that they must contain much more mass than is present in the form of luminous matter. There is no shortage of evidence in favor of dark matter.

That statement carries an important caveat, however. Although a wide variety of observations suggests that dark matter exists, each of these measurements infers the presence of dark matter through the effects of its gravity. Our theory of gravity predicts that without dark matter, galaxies would not rotate the way they do. Our theory of gravity predicts that dark matter pulled ordinary matter together into clumps to create the structural imprints we see in the cosmic microwave background. Our measurements of dwarf galaxies and galaxy clusters enable us to infer the presence of dark matter, again through the effects of gravitation. Every piece of evidence we have for the existence of dark matter is based on its expected gravitational interaction with the luminous matter of our world. Hearing this, a skeptic might ask, "Instead of dark matter existing, might it be that we have misunderstood the force of gravity? Instead of an extra kind of matter, might there be an extra kind of gravity?" These are good questions.

The modern theory of gravity, Einstein's general theory of relativity, has been incredibly successful, passing every experimental test that it has undergone. But now the skeptic steps in. He asks, "But what kind of tests were performed?" Once again, a good question. Naturally, these tests have their limitations. For one, although the force of gravity can be tested with incredible accuracy over modest distances—over the size of our solar system, for example—it is far more difficult to make equally precise measurements over much greater distances. Intriguingly, it is over such larger distances—which correspond to the size of galaxies and clusters of galaxies—that the effects of dark matter have been detected. On top of this, the gravitational acceleration produced from dark matter can be very, very small. Just how moving bodies behave under the conditions of such tiny acceleration has also not been precisely tested.

In 1983 the Israeli physicist Mordechai Milgrom first proposed that the observed behavior of rotating galaxies, instead of being evidence for the existence of dark matter, might be evidence for a new feature of gravity. In his original paper on the subject, Milgrom showed that by changing the laws of Newtonian physics under the conditions of very small accelerations, the observed rotation rates of galaxies could be matched without dark matter. Because the tests of the force of gravity at very small accelerations are so difficult to perform with much accuracy, these modifications to Newton's theory could have gone undetected. This new theory of Milgrom's became known as modified Newtonian dynamics, or MOND for short.[2]

Milgrom and others to follow made a number of predictions with the theory of MOND. By and large, those predictions seem to be in reasonably good agreement with subsequent observations. More than a hundred galaxies have had their rotations studied and compared to the expectations of MOND, often with favorable results. Milgrom's paper also predicted that in a class of dim galaxies, called low-surface-brightness galaxies, the effects of MOND would be especially pronounced. Observations seem to support this conclusion as well. MOND has been quite successful in predicting and explaining the behavior of galactic rotations without the existence of dark matter.

Despite MOND's successes, not all of the theory's predictions have agreed so well with observations. One particularly important problem facing MOND is that the total amount of mass present in large clusters of galaxies appears to be substantially greater than the quantity of ordinary, luminous matter, even if the effects of MOND are taken into account. This disagreement has been seen by some as the fatal failure of MOND and as a reaffirmation of the existence of dark matter. This issue is not yet resolved. Although probably unlikely, it is possible that the observational esti-

---

2. Note that Milgrom's theory of MOND is a modification of Newton's theory of gravity rather than Einstein's general theory of relativity. This means that MOND cannot be used reliably under all conditions. This is not a problem for calculating the rotations of galaxies, since the theories of Newton and Einstein are in good agreement with each other in that application. The same is not true when addressing questions of cosmology, however.

mates of the masses of galaxy clusters might be refined further in the future—maybe even enough to be compatible with the theory of MOND.

In addition to the problem of galaxy clusters, modern cosmological observations have also proven to be a challenge to MOND. Cosmological measurements, such as those of the cosmic microwave background, have become increasingly precise over the past several years. These observations have continued to agree very well with what is expected for a universe containing a great deal of dark matter. A successful MOND theory would also have to be compatible with these new pieces of cosmological data.

Until very recently it wasn't clear whether any kind of MOND theory would be consistent with these modern cosmological measurements. It wasn't clear because Milgrom's paper and the other MOND papers that followed didn't describe a complete theory. In particular, MOND was not fully compatible with Einstein's general relativity, and so it could not be used to make predictions under all circumstances. Because it is general relativity, and not Newton's theory, that describes how the Universe expands and evolves, Milgrom's MOND can't be applied to these phenomena. As only a "toy model," MOND was of little use in cosmology.

The situation changed in 2004 when Jacob Bekenstein, a Mexican physicist working in Israel, proposed a new version of MOND. Unlike Milgrom's theory, Bekenstein's MOND was formulated within the context of general relativity. Instead of being a modification of Newtonian physics, Bekenstein's theory is a modification of Einstein's relativity. With this new theory in hand, one could finally begin to test MOND against cosmological data.

The first, and as of the time I am writing the only, study of the cosmological implications of Bekenstein's version of MOND was recently carried out at Oxford University. In the interest of transparency, I must say that I held a position at Oxford while this study was being conducted, and that I am friends with the physicists—Constantinos Skordis, Pedro Ferreira, David Mota, and Celine Boehm[3]—who carried it out. My friendship with them aside, I think I can say without bias that the calculation they performed was no walk in the park. Bekenstein's theory of MOND is notoriously complicated and difficult to manipulate. Several times in this book, I have mentioned the difficulty involved in finding solutions to the equations of Einstein's general theory of relativity. Well, Bekenstein's theory is essentially an extra-complicated version of Einstein's theory. To make much headway in understanding it requires a highly methodical and thorough approach. Fortunately, this is exactly the scientific style of Constantinos Skordis. Compared to him, I conduct research about as methodically as the Cat in the Hat cleans house.

The Oxford group sought to determine whether the features observed in the cosmic microwave background were consistent with Bekenstein's theory. When they published their results in the spring of 2005, their answer was a *yes,* with a *but.* Yes, the observed features could be produced within the framework of Bekenstein's model, but not without an extra amount of matter thrown into the mix.[4] When I first heard of this conclusion, I immediately asked, "If you have to add extra matter in addition to the new behavior of MOND, wouldn't it be easier just to add even more matter—dark matter—instead of MOND?" After all, adding extra matter that we don't detect into the Universe is adding dark matter. Perhaps it is not as much dark matter as is needed without MOND, but it is some nonetheless.

As I see it, there are two major problems facing the theory of

3. You might recall Celine from chapter 6.

4. The Oxford group's paper used neutrinos for this extra matter.

MOND today. First are the observations of galaxy clusters and MOND's inability to explain them. Second are the challenges involved in testing MOND against the precision cosmological observations that have been made, particularly those of the cosmic microwave background. So far, only Bekenstein's version of MOND can make cosmological predictions, and it has been shown to require at least some dark matter for it to agree with the current observations.

Despite these challenges, it may still be possible that future observations and studies will lead to a way of reconciling the measurements of galaxy clusters with the predictions of MOND, or that some other formulation of MOND will be found that is consistent with modern observational cosmology without the presence of dark matter. With this possibility remaining, MOND as an alternative to dark matter is not a dead hypothesis. After the beating it has taken from the observations of galaxy clusters and the cosmic microwave background, however, it appears to be badly wounded and on life support.

• • •

It has been more than seventy years since Fritz Zwicky first proposed dark matter, and more than fifty since Vera Rubin published her first findings supporting it—plenty of time for the dark matter hypothesis to be tested, challenged, and largely accepted, even if skeptics remain. The existence of dark energy, on the other hand, is a relatively new realization. Prior to 1998, although there was some evidence that the total density of our Universe was larger than the total matter density, few cosmologists took this as serious evidence for the presence of dark energy. It wasn't until 1998 that the Supernova Cosmology Project and the High-Z Supernova Search Team first announced their discovery that the Universe was expanding at an accelerating rate. It was only with this observation that most cosmologists became convinced that dark energy exists. Dark energy, as a concept, is still in its infancy.

Along with the discovery of dark energy, skepticism regarding dark energy's existence is in its infancy as well. Just as it takes time to develop experiments and theories that support a scientific conclusion, it takes time to poke holes in the evidence of a discovery, as well as to propose and develop alternative explanations for the evidence. Although there is a healthy degree of skepticism regarding dark energy in the cosmological community today, the alternatives are really only beginning to be explored.

Before arriving at Oxford in 2003, I had never given the evidence for dark energy much thought. The scientific community had largely endorsed the conclusion that we live in a Universe dominated by dark energy, and I, a new researcher who had earned his PhD a mere few months earlier, didn't see much reason to doubt that they were right. While at Oxford, however, I met and became a friend and collaborator of a physicist named Subir Sarkar. Along with being a thoroughly detailed and rigorous particle physicist and cosmologist, Subir is also a dark-energy skeptic.

As the evidence for dark energy accumulated, it did not take long for something of a consensus to be reached by the cosmological community. Subir was one of the cosmologists who paused long enough to ask what assumptions this conclusion was based on. Further, he wondered whether other reasonable assumptions could be adopted that would not lead to the conclusion that dark energy exists. Subir assessed the evidence for dark energy in much greater detail than most other cosmologists had. In the end, Subir wasn't entirely convinced that dark energy exists.

During the two years I spent at Oxford, Subir did not manage to convince me that dark energy is not likely to exist. Nor do I think that was his intention. He did, however, convince me that the evidence in favor of dark energy might not be entirely conclusive. Sometimes the job of a skeptic is not to tell you what he knows, but to tell you what is still unknown.

Although the criticisms of the evidence for dark energy are varied, many of these objections hinge on questions regarding the reliability of using supernovae to measure the expansion rate of our Universe, and thus challenge the assertion that the expansion rate is accelerating. Recall from chapter 9 that all type Ia supernovae have approximately the same intrinsic brightness, and thus can be used as "standard candles" enabling us to tell how fast distant objects are moving away from us. With that data, we can map out the expansion history of our Universe, and come to the conclusion that dark energy exists.

But a skeptic asks how much we can trust this conclusion. It could be skewed by a poor understanding of type Ia supernovae: Perhaps those explosions had different properties billions of years ago that make more distant—and older—supernovae appear to be less bright than we expect them to be. Perhaps some yet-unknown process is affecting the light traveling from a distant supernova to Earth. Or perhaps type Ia supernovae are not as standard as we think.

Although we don't have much reason to doubt the conclusions drawn from distant supernova measurements, I can't rebut the criticisms regarding these observations entirely. To be certain that dark energy exists, other kinds of tests are needed to confirm these results. Other cosmological measurements are essential to support the conclusion that dark energy exists.

Fortunately, we do have such independent data. Our studies of the cosmic microwave background have demonstrated that the

Universe is roughly flat, with a total density approximately equal to the critical density. In contrast, when we measure the density of matter in our Universe by observing clusters of galaxies, we find that the density is much smaller: about one-third of the critical density. Thus, even if we completely disregard the observations of distant supernovae, other cosmological measurements lead us to the conclusion that our Universe must contain much more than just matter. Dark energy, it seems, has left its fingerprints in more than one place.

But there is a catch. Those conclusions rely on a set of assumptions, as all scientific findings do. Some studies adopt assumptions that are not at all controversial or dubious, and therefore they produce findings that are rarely challenged. Others are built upon assumptions that are less certain. The skeptic asks how certain we are about the assumptions that underlie our belief in dark energy.

The trick in answering this question is that a flawed assumption often looks like a safe assumption until a problem with it is brought to light. So far, none of the assumptions leading to the conclusion that dark energy exists have been shown to be false. Nevertheless, Subir Sarkar and his collaborators have been exploring the possibility that the technical assumptions used in the analysis of the cosmic microwave background might not be reliable. If this is true, then the cosmic microwave background might not be capable of conclusively demonstrating that dark energy exists. Although we cannot be sure that this will not eventually be found to be the case, I find it rather unlikely—increasingly so as more and better observations are made. And even if the conclusions drawn from the cosmic microwave background turn out to be inconclusive, we would still need to find a separate explanation for why the observations of distant supernovae appear to show that the Universe's expansion rate is accelerating. Dark energy appears likely to be with us to stay.

• • •

The one sure way to dispel scientific skepticism is to produce irrefutable evidence.[5] In the case of dark matter, bulletproof evidence might come in the form of a direct discovery of dark matter particles. This could very plausibly occur over the next few years in a particle collider, or in direct or indirect dark matter detection experiments like those I discussed in chapter 6. Were we to "see" dark matter in the laboratory, it would be much harder to argue that it does not also exist throughout the Universe. Dark energy, on the other hand, is almost certainly more difficult than dark matter to detect in a laboratory. But there are other ways to make a bulletproof case. Future cosmological measurements will search for the effects of dark energy in more types of data, I hope settling this issue once and for all.

Over the next several years, new experiments will likely reveal to us a great deal about dark matter and dark energy. It might be that the new observations will support the current mainstream conclusions, leading many of the skeptics to concede the argument. Scientific history is replete with such episodes, from the acceptance of the Sun-centered solar system to the verification of relativity and quantum physics. If, on the other hand, results are found that conflict with the present consensus, more members of the scientific community will begin to consider alternative expla-

---

5. The key word in this sentence is "scientific." Irrefutable evidence for biological evolution has been available for quite some time, and yet skepticism abounds. It would seem that unscientific skepticism is rarely concerned with evidence.

nations. Today's outsider skeptics may someday find themselves spokesmen for the mainstream of science.

During recent years, the evidence in support of dark matter and dark energy has been growing. Despite this apparent trend, nobody can predict where the coming years and decades of cosmology might lead us. The future of science is always clouded, and never certain.

CHAPTER 13

VISIONS OF THE FUTURE

Prediction is very difficult, especially about the future.

—Niels Bohr

I never think of the future—it comes soon enough.

—Albert Einstein

Predictions of the future often fail. Flying cars have yet to become commonplace, and no one lives on the Moon. Sometimes, however, predictions do come more or less true. H. G. Wells foresaw powerful bombs based on radioactive elements, even if he had no idea how they would work. That, really, is the key. Futurologists, science fiction writers, and other modern-day prophets sometimes get new technologies approximately right, but they have invariably missed the events of greatest significance—revolutions in thinking that enable us to make dreamed-of technology work. The same is true of scientists. No one in Newton's time foresaw the theory of relativity or quantum physics. Nineteenth-century physicists were merely trying to confirm the presence of the ether when they started down the path that led to the concept's abandonment and the dual revolutions in physics led by Einstein, Planck, and Bohr. I expect that, just as those physicists could not have imagined the discoveries that lay before them, the physicists of today are equally in the dark. That said, I will now proceed with my own predictions, written from the vantage point of the next century, looking back on the history of cosmology. I'll admit in advance that the entry below quickly moves from the historical to rank speculation to the most reckless fantasizing.

From the *Encyclopedia Britannica* (2100 Edition)
Cosmology (history of):

The history of cosmology as a branch of science began in the early twentieth century with the development of Albert Einstein's general theory of relativity, first published in 1915. In the years following, numerous solutions to Einstein's equations were found, describing how a universe evolves with time. Einstein himself manipulated the equations to describe a static universe. In 1929, however, American astronomer Edwin Hubble observationally determined that the Universe is expanding, leading to the conclusion that the Universe is not static but had evolved over billions of years from a highly dense and hot state.

That conclusion, which came to be known as the Big Bang theory, became the accepted description of the evolution of the Universe. Among the evidence supporting this theory were the relative abundances of hydrogen, helium, and other light chemical elements present in the Universe, and the observation of the cosmic microwave background, each of which was accurately predicted by the Big Bang theory.

In the 1990s detailed observations of the cosmic microwave background were used to determine that the Universe was geometrically flat, as described by the traditional Euclidean laws of geometry. That determination provided support for the hypothesis that a period of ultra-fast expansion, called inflation, took place in the very early Universe. Further studies of the cosmic microwave background, made in the early twenty-first century, along with the observation of gravitational waves in the 2030s, confirmed that an inflationary era had in fact taken place.

Observations made in the 1990s of unexpectedly dim, distant supernovae suggested that the expansion rate of the Universe was accelerating. That observation, along with other supporting data, led to the conclusion that the majority of the Universe's energy density—some 70 percent—existed in a form then called dark energy.

Astronomers had known since the 1970s that most of the Universe's matter density existed in an invisible form called dark matter. The nature of this material was not identified, however, until 2010, when a particle accelerator in Switzerland, called the Large Hadron Collider, discovered a number of new elementary particles. Many of these new types of matter were determined to be those predicted by the theory of supersymmetry; one of those particles, the neutralino, had long been considered the most promising candidate for dark matter. That the dark matter is in fact made up of neutralinos was confirmed in

2016 through observations of energetic neutrinos, gamma rays, and positrons coming from space. In 2020 neutralino dark matter particles were observed directly for the first time in deep underground detectors.[1]

The nature of dark energy proved to be more difficult to determine than that of dark matter, remaining a mystery for nearly eighty years after its discovery. Only after the development of the Kohler-Stravas theory in the 2070s, which unified Einstein's theory of general relativity with the theory of quantum physics, was any substantial progress made in this area. That theory, a descendant of the string theories of the late twentieth and early twenty-first centuries, predicted the presence of an energy density associated with specific types of vibrating particle-gravity strings. That energy has since been found to be the dark energy first discovered in the late twentieth century.[2]

In 2091 the science of cosmology underwent what is generally acknowledged to be its greatest development. This discovery, known as the Juavis revolution, arose from a series of experiments that . . . [3]

Finally heeding my own warning about the futility of prediction, I stop here. I might be speculating reasonably about the identity of dark matter, about the development of a theory of quantum gravity, or that such a theory could lead us to an understanding of the nature of dark energy. But it is a fool's errand to predict the nature of a revolution.

1. This paragraph is wild speculation.

2. This paragraph is ridiculously wild speculation.

3. What is the word that means more ridiculous than ridiculous?

• • •

As I write this chapter, I see such potential for scientific discovery along the horizon that sometimes I have a hard time imagining the next decade passing without major discoveries being made in particle physics, cosmology, or both. Some of my more experienced colleagues tell me that I manage to maintain this viewpoint only because I am still a young scientist who hasn't yet been jaded by a series of anticipated discoveries that never happened or by other disappointments. Perhaps these naysayers are right. After all, thirty years ago it was common for particle physicists to think they were only a few years away from uncovering a grand unified theory. If I had been a young researcher in the late 1970s instead of now, I would probably have found myself quite disappointed.

Nevertheless, when I try to make a quasi-objective assessment of the future of cosmology and particle physics, I can't help but be excited. The experiments currently being developed, constructed, and deployed have incredible prospects. Of them all, the one with the greatest chance of discovery is one I cited in my encyclopedia of the future—the Large Hadron Collider.

The Large Hadron Collider, or LHC, will accelerate beams of protons through a circular tunnel some seventeen miles long buried about a hundred yards beneath the city of Geneva, Switzerland. Two streams of protons, traveling in opposite directions at about 99.999999 percent of the speed of light, will collide head-on. Each collision will contain up to 14 trillion electron volts, or 14 thousand giga-electron volts, of energy. So far, the most energetic collisions studied at a particle accelerator, at the Tevatron at the Fermi National Accelerator Laboratory, have had less than two thousand giga-electron volts. The Tevatron is an impressive machine, but the LHC is a big step forward.

The amount of energy an accelerator can reach is directly related to the types of particles an experiment can study or (potentially) discover. Einstein's relationship, $E=mc^2$, implies that to produce a particle with a particular mass, you need an equivalent amount of energy. So far, the most massive particle yet discovered, the top quark, has mass equivalent to about 175 giga-electron volts of energy. The second- and third-most massive particles yet discovered are the W and Z bosons, which have masses of about 80 and 90 giga-electron volts. Other than those three, all of the known particles have masses no greater than a few giga-electron volts. We are quite certain, however, that these are not the only kinds of particles that exist in our Universe. Assuming that supersymmetry exists in nature, there should be a slew of new particles with masses between roughly a few tens and a few thousands of giga-electron volts—perfectly suited for discovery at the LHC. And even if supersymmetry does not exist, it is very likely that something with a mass in that range does. Our current collection of known particles just isn't a complete picture. It's hard to imagine the LHC not discovering something new once it begins operation.

Particle accelerators are not the only avenue for discovery in particle physics. Underground detectors, enormous neutrino telescopes, gamma-ray- and antimatter-detecting satellites, and other experiments should collectively either confirm or rule out many of the most popular dark-matter candidates.

And cosmology seems ripe for further discovery, too. Improving upon COBE and WMAP, a satellite experiment called Planck stands to make the highest-quality maps of the cosmic microwave background yet. More detailed supernova surveys will improve our measurements of the Universe's expansion rate over its history and (I hope) move us closer to discerning the true nature of dark energy.

• • •

With all of these experiments soon to enlighten us, I feel entitled to my optimism. But just for the sake of evenhandedness, let me put on my pessimist's hat, at least for a moment or two. With that hat tightly pulled down on my head, I can imagine a second, alternative twenty-second-century encyclopedia entry that ends more like this:

> . . . These observations led a large fraction of the scientific community to conclude that the majority of the Universe's energy density exists in the form of undetected matter and energy, the so-called dark matter and dark energy. Within the scientific community, this dark matter was commonly expected to consist of weakly interacting massive particles. This consensus was undermined, however, when experiments such as the Large Hadron Collider revealed no such particles. Although the Large Hadron Collider, and the International Linear Collider that followed, did discover a large number of new particles during their operation, including many of the "superpartners" predicted by the theory of supersymmetry, none were sufficiently stable to constitute a candidate for dark matter.
>
> This revelation, among others, led to the abandonment of the hypothesis that dark matter consists of weakly interacting particles. Since then, the leading candidates for dark matter have been elementary particles that interact only through the force of gravity, making them extremely difficult to detect directly.
>
> The effort to identify the nature of dark energy has made similarly poor headway since its discovery in the

late twentieth century. Although its presence has been confirmed through a variety of measurements, dark energy remains very poorly understood today.

I think of this outcome as the worst-case scenario. I find it thoroughly depressing to imagine spending years or decades of my career searching for something that is practically impossible to find. Fortunately, this scenario seems unlikely to me. Even if dark matter interacts only through gravity, it is possible that it could be studied in a collider experiment by observing other, heavier particles decay into the dark matter particles. Similarly, it is possible that no big break will come soon in unraveling the puzzle of dark energy, but I doubt it.

Thinking more generally, I have confidence in the ingenuity of many of the physicists working on these puzzles. If the searches for dark matter we are conducting now don't find anything, new means of investigation will be developed. I have many very smart and inventive colleagues. There are not many problems that they, collectively, cannot find a solution to.

• • •

It's very possible that none of the scenarios I have imagined actually come close to what will someday be revealed as the truth about dark matter and dark energy. As my former PhD adviser, Francis Halzen, once told me, "The probability of any of the things you have thought of being true is almost always smaller than the probability of the sum of all of the things you haven't thought of." And, to be honest, that's okay with me. Discovery is what makes science exciting. Any time a new piece of a puzzle is seen for the first time, it is possible that the whole picture might suddenly become clear before your eyes. Just like filling in one word in a crossword puzzle can make the next word and the next after it suddenly become obvious, scientific discoveries often beget more discoveries. Could the discovery of dark matter's identity start a domino effect of scientific progress? Could it lead to the discovery of a grand unified theory, or the understanding of dark energy, or a Theory of Everything? I don't know. No one can tell where an unanticipated discovery might lead. I just hope that someday I get to see it happen.

INDEX

Page numbers in *italics* refer to illustrations.